Springer
Tokyo
Berlin
Heidelberg
New York
Barcelona
Hong Kong
London
Milan
Paris

Y. Baba, A.J. Hayter
K. Kanefuji, S. Kuriki (Eds.)

Recent Advances in Statistical Research and Data Analysis

With 14 Figures and 31 Tables

 Springer

Yasumasa Baba
Professor
Center for Information on Statistical Sciences
The Institute of Statistical Mathematics
4-6-7 Minami Azabu, Minato-ku, Tokyo 106-8569, Japan

Anthony J. Hayter
Associate Professor
School of Industrial and Systems Engineering
Georgia Institute of Technology
Atlanta, GA 30332-0205, USA

Koji Kanefuji
Associate Professor
Center for Information on Statistical Sciences
The Institute of Statistical Mathematics
4-6-7 Minami Azabu, Minato-ku, Tokyo 106-8569, Japan

Satoshi Kuriki
Associate Professor
Department of Fundamental Statistical Theory
The Institute of Statistical Mathematics
4-6-7 Minami Azabu, Minato-ku, Tokyo 106-8569, Japan

ISBN 978-4-431-68546-3   ISBN 978-4-431-68544-9 (eBook)
DOI 10.1007/978-4-431-68544-9

Library of Congress Cataloging-in-Publication Data applied for.

Printed on acid-free paper

Typesetting: Editors and authors

SPIN: 10831869

Preface

The Institute of Statistical Mathematics (ISM), founded in 1944, is a research institute of the Ministry of Education, Culture, Sports, Science and Technology (MEXT). In 1997, ISM restructured the Center for Development of Statistical Computing (CDSC) and the Center for Information on Statistical Sciences (CISS), and at the same time created visiting professorships for researchers. In addition, ISM now accepts researchers from around the world through programs funded by MEXT and the Japan Society for the Promotion of Science.

ISM symposia have been planned by CISS to promote collaboration between visiting researchers and Japanese researchers. Since the first ISM Symposium was held in 1999, CISS has organized the following meetings:

> Data Mining and Knowledge Discovery in Data Science (March 18–19, 1999)

> Recent Advances in Statistical Research and Data Analysis (March 21–22, 2000)

> Statistical Software in the Internet Age (February 19–21, 2001)

> Statistical Researches in Complex Systems (March 8–9, 2001)

This book is a collection of papers presented at the ISM Symposium on Recent Advances in Statistical Research and Data Analysis, held in March 2000. Anthony J. Hayter and Shizuhiko Nishisato were at ISM as visiting professors in 1998. Prof. Hayter returned to ISM as a visiting associate professor in 1999 and 2000, and during his second visit, the symposium was held, thanks in part to his proposal for a symposium to cover a wide range of topics in statistics, from theory to data analysis. He delivered the keynote address, and six special lectures were presented, by Jorge J. Riera, Chihiro Hirotsu, Shizuhiko Nishisato, Myoungshic Jhun, Tatsuo Otsu, and Myung-Hoe Huh. In addition there were twelve contributed papers.

This volume, *Recent Advances in Statistical Research and Data Analysis*, contains the keynote speech and four of the special lectures.

Because all the papers are concerned with theory and methods for real data, we believe that the book will benefit researchers, students and those engaged in data analysis.

We express our cordial gratitude to Dr. Ryoichi Shimizu, Director General of ISM, for his generous support which made the publication of this book a reality.

Tokyo, January, 2002   Yasumasa Baba

Contents

Multiplicity Problems in the Clinical Trial and Some Statistical Approaches

Chihiro Hirotsu

Graduate School of Engineering, University of Tokyo
Bunkyo-Ku, Tokyo 113-8656, Japan
hirotsu@stat.t.u-tokyo.ac.jp

Summary: In a clinical trial there arise a variety of multiplicity problems which might cause the bias and need to be considered carefully in analyzing and interpreting the data. We first introduce those problems frequently encountered in practice and then discuss some statistical approaches to overcome the difficulty.

1. Introduction

In the clinical trial various types of multiplicity problems have been discussed, see, for example, Armitage and Parmar (1986). They include the problems of

1. multiple endpoints,
2. subgroup analysis,
3. different methods of analyses simultaneously applied to one set of data,
4. multiple comparisons of treatments,
5. repeated χ^2 tests for a $2 \times K$ contingency table with ordered column categories,
6. repeated measurements,
7. interim analyses.

The first two of these problems have been argued as most awkward by Armitage and Parmar. After long discussions, however, the first problem has been settled by stating clearly in the protocol very few primary endpoints of the trial, see Lewis (1999), for example. Regarding the subgroup analysis the temptation is to select a posteriori subgroups with apparently large treatment effects or to assert an interaction to exist between the treatment and a particular form of classification such as sex or age. It is now strongly recommended to

1

describe in the protocol what sort of interaction is expected to exist. Otherwise the observed interaction is required to be reconfirmed by another trial. Regarding the third problem it is now requested to describe the detail of the method of analysis in the protocol at least for the primary endpoint. In the present paper we therefore address the other problems where more or less statistical approaches are available.

In Section 2 some varieties of methods for multiple comparisons of treatments are introduced with particular interest in the case where there is a natural ordering in the treatment. They include the maximal t test which is extended in Section 3 to comparing ordered binomial probabilities and called the maximal chi-squared test. In section 4 it is extended to the analysis of two-way contingency table with ordered row and/or column categories and in Section 5 to the analysis of association between the disease and allele. Then the idea is extended to the haplotype analysis of three-way association between the disease and bivariate allele frequencies at two closely linked loci. As a by-product a factorization of the probability distribution of the three-way cell frequencies is obtained under the null hypothesis of no three-way interaction. In Section 6 a row-wise multiple comparisons procedure is introduced for a two-way layout model with ordered column categories with an example of application to comparing treatments based on repeated measures.

2. Varieties of Methods for Multiple Comparisons of Treatments

The most well known multiple comparisons procedure of several treatments will be Tukey and Scheffé methods. The most popular in the clinical trial will, however, be the Dunnett (1964) test comparing several treatments with a control. Also well known are the Williams (1971) test and its modification by Marcus (1976) for comparing ordered treatments in the dose-response analysis which is based on the order restricted maximum likelihood estimators (mle). Although these methods are useful for the monotone hypothesis of treatment effects in a one-way layout, a difficulty arises in calculating the mle and its cumulative distribution function under the more complicated situations such as convexity, sigmoidicity and further two-way ordered hypotheses beyond the one-way layout setting. Here we introduce the max t test for the ordered hypothesis which is not based on the mle and easily extended to those more complicated situations. These include the analysis of contingency tables as shown in Sections 3, 4 and 5, see also Hirotsu (1993).

Consider a one-way layout model

$$y_{ij} = \mu_i + \varepsilon_{ij}, \quad i = 1, \ldots, K; \ j = 1, \ldots, n_i,$$

where ε_{ij} is assumed to be independently and identically distributed as $N(0, \sigma^2)$. Now, suppose we are interested in testing the null hypothesis

$$H_0 : \mu_1 = \cdots = \mu_K$$

against an ordered alternative

$$H_1 : \mu_1 \leq \cdots \leq \mu_K \text{ with at least one inequality strict.}$$

Let $\bar{\boldsymbol{y}}$ be a vector of the averages $\bar{y}_{i\cdot}$. Then the max t is the maximal standardized element of a vector

$$\boldsymbol{t} = (D_K' D_K)^{-1} D_K' \text{diag}(n_i)(\bar{\boldsymbol{y}} - \bar{y}\boldsymbol{j}),$$

where $\bar{y}$ is an overall average of observations, $\boldsymbol{j}$ a vector of 1's,

$$D_K' = \begin{bmatrix} -1 & 1 & 0 & \cdots & 0 & 0 \\ 0 & -1 & 1 & \cdots & 0 & 0 \\ \multicolumn{6}{c}{\dotfill} \\ 0 & 0 & 0 & \cdots & -1 & 1 \end{bmatrix}_{K-1 \times K}$$

and

$$(D_K' D_K)^{-1} D_K' = \frac{1}{K} \begin{bmatrix} -(K-1) & 1 & 1 & \cdots & 1 & 1 \\ -(K-2) & -(K-2) & 2 & \cdots & 2 & 2 \\ \multicolumn{6}{c}{\dotfill} \\ -1 & -1 & -1 & \cdots & -1 & (K-1) \end{bmatrix}.$$

The columns of $D_K(D_K' D_K)^{-1}$ define a set of corner vectors of a convex cone defined by H_1 and the max t is known to be the standardized maximum of the projections of the observations on to those corner vectors, see Hirotsu (1997). It is interesting to note that an essentially complete class of tests for H_0 against H_1 is given by all those tests that are increasing in every element of the vector $\boldsymbol{t}$ and with a convex acceptance region (Hirotsu, 1982). Very interestingly the columns of $D_K(D'_K D_K)^{-1}$ define the components of a changepoint model

$$K_1 : \mu_1 = \cdots = \mu_\tau < \mu_{\tau+1} = \cdots = \mu_K,$$

where $\tau = 1, \ldots, K - 1$ are the unknown changepoints and the $\max t$ is characterized also as the likelihood ratio test for K_1. The derivation of $\max t$ as a likelihood ratio test against the changepoint hypothesis K_1 is similar to that given by Hawkins (1977) although the two-sided version

$$K_2 : \quad \mu_1 = \cdots = \mu_\tau < \mu_{\tau+1} = \cdots = \mu_K$$
$$\text{or}$$
$$\mu_1 = \cdots = \mu_\tau > \mu_{\tau+1} = \cdots = \mu_K$$

is considered there. One of the most eminent features of t is that the subsequent components of it satisfy the first order Markov property which leads to an exact and very efficient algorithm for calculating the p value of the $\max t$. We briefly mention the algorithm below.

Let the standardized components of t be denoted by $(t_1, \ldots, t_{K-1})$ and assume the variance σ^2 is known. It should be noted that the kth component t_k is the uniformly most powerful test against the changepoint hypothesis between the k and $k + 1$st point and given by

$$t_k = \left\{ \frac{\sigma^2}{N_k} + \frac{\sigma^2}{N_k^*} \right\}^{-1/2} \left(\frac{Y_k^*}{N_k^*} - \frac{Y_k}{N_k} \right), \quad k = 1, \ldots, K - 1, \qquad (2.1)$$

where

$$Y_k = y_{1\cdot} + \cdots + y_{k\cdot}, \quad Y_k^* = y_{k+1\cdot} + \cdots + y_{K\cdot},$$

$$N_k = n_1 + \cdots + n_k, \quad N_k^* = n_{k+1} + \cdots + n_K,$$

with $y_{i\cdot}$ the total observation at the ith level. When σ^2 is unknown we replace it by the usual unbiased estimator of the variance

$$\hat{\sigma}^2 = \frac{\sum_i \sum_j (y_{ij} - \bar{y}_{i\cdot})^2}{N_K - K}, \qquad (2.2)$$

and then t_k is simply the t test comparing the averages of means

$$\sum_{k+1}^{K} n_i \mu_i / N_k^* \quad \text{with} \quad \sum_{1}^{k} n_i \mu_i / N_k.$$

In the following we assume σ^2 known but the unknown case can be treated similarly. Now we are interested in calculating the distribution function

$$F(t) = \Pr\{t_1 \leq t, \ldots, t_{K-1} \leq t\}.$$

For the purpose define a conditional probability

$$F_k(t|t_k) = \Pr\{t_1 \leq t, \ldots, t_k \leq t|t_k\}$$

given t_k. Define also an initial function

$$F_1(t|t_1) = \begin{cases} 1, & t_1 \leq t, \\ 0, & \text{otherwise.} \end{cases}$$

Then we have following Lemma.

Lemma 1.   Recurrence formula 1

$$F_{k+1}(t|t_{k+1}) = \begin{cases} \int_{-\infty}^{t} F_k(t|t_k) f(t_k|t_{k+1}) dt_k, & t_{k+1} \leq t, \\ 0, & \text{otherwise,} \end{cases} \qquad (2.3)$$

where $f(t_k|t_{k+1})$ is the conditional distribution of t_k given t_{k+1}.

Proof of Lemma 1.   By the law of total probability we can rewrite the $k+1$st conditional probability as

$$F_{k+1}(t|t_{k+1}) = \int \Pr\{t_1 \leq t, \ldots, t_k \leq t, t_{k+1} \leq t|t_k, t_{k+1}\} f(t_k|t_{k+1}) dt_k.$$

If $t_{k+1} \leq t$ then we have

$$\begin{aligned} F_{k+1}(t|t_{k+1}) &= \int \Pr\{t_1 \leq t, \ldots, t_k \leq t|t_k, t_{k+1}\} f(t_k|t_{k+1}) dt_k \\ &= \int \Pr\{t_1 \leq t, \ldots, t_k \leq t|t_k\} f(t_k|t_{k+1}) dt_k \\ &= \int F_k(t|t_k) f(t_k|t_{k+1}) dt_k. \end{aligned}$$

The second equality comes from the Markov property. If $t_{k+1} > t$ then $F_{k+1}(t|t_{k+1})$ is obviously 0.

   We extend the definition of F_k up to $k = K$ assuming t_K to be a sufficiently small constant. Then we have $F(t)$ as $F_K(t|t_K)$, where $f(t_{K-1}|t_K) = f(t_{K-1})$ is an unconditional probability density function. Finally the p value for the observed $t_0 = \max t$ is given by $p = 1 - F(t_0)$.

Table 1: Half-life of an anti-biotics in rats

Dose level	(mg/Kg/Day)	Data					t_k
1	(5)	1.17	1.12	1.07	0.98	1.04	6.17
2	(10)	1.00	1.21	1.24	1.14	1.34	8.58
3	(25)	1.55	1.63	1.49	1.53		7.20
4	(50)	1.21	1.63	1.37	1.50	1.81	7.44
5	(200)	1.78	1.93	1.80	2.07	1.70	

Example 1.   We apply the method to the half-life data of an anti-biotics in Table 1 obtained from a dose-response experiment in rats, where the statistics t_k are given in the last column. In this case σ^2 is unknown and we used the $\hat{\sigma}^2$ of (2.2).

In this case the maximal t value 8.58 is obtained between the dose levels 2 and 3. The exact p value of it is evaluated as 1.1×10^{-7} by the recurrence formula (2.3) modified accommodating to the σ unknown case. We can apply the closed testing procedure by Marcus et al. (1970) by simply cutting off the highest dose and applying the same procedure repeatedly until nonsignificant result is obtained. Then at the 2nd and 3rd steps we obtain the maximal t values 5.95 and 4.92 which are highly significant. At the 4th step we obtain t value 1.21 as the simple two sample t statistic for comparing the dose levels 1 and 2, which is not significant at significance level 0.05. Thus we conclude that the responses for the higher dose levels 3, 4, 5 are significantly different from those for the lower dose levels 1 and 2.

We can derive extended $maxt$ tests which posses similar characteristics with $\max t$ against the convexity

$$H_2 : \mu_2 - \mu_1 \leq \mu_3 - \mu_2 \leq \cdots \leq \mu_K - \mu_{K-1}$$

and the sigmoidicity hypothesis

$$H_3 : \mu_1 - 2\mu_2 + \mu_3 \geq \mu_2 - 2\mu_3 + \mu_4 \geq \mu_{K-2} - 2\mu_{k-1} + \mu_K,$$

respectively. The corner vectors of the convex cone defined by H_2 define the slope change model and those of H_3 define an inflection point model. A very efficient algorithm for the distribution function of the extended $\max t$ is again obtained, which is based on the second and the third order Markov properties of the subsequent components of the test statistics, see Hirotsu and Marumo (2002) for details.

Table 2: Data from a dose-response experiment

Dose level	Improved	Not improved	Total
1	20	16	36
2	23	18	41
3	27	9	36
4	26	9	35
5	9	5	14

The analogue of $\max t$ for a contingency table has been called as $\max \chi^2$ and some applications of the statistic are mentioned in the following sections.

3. Comparing Ordered Binomial Probabilities

The data in Table 2 are from a dose-response experiment carried out as a phase II clinical trial for 5 dose levels of a drug for heart disease. For comparing these dose levels the analogue of $\max t$ is derived.

Denote the number of improved patients by y_i and the total number of observations by n_i for the ith dose level. We assume a binomial distribution $B(n_i, p_i)$ at the ith level and our interest is in testing the null hypothesis

$$H_0 : p_1 = \cdots = p_K$$

against an ordered alternative

$$H_1 : p_1 \leq \cdots \leq p_K$$

with at least one inequality strict. Then the analogue of $\max t$ is derived as follows.

Define the accumulated statistic

$$Y_k = y_1 + \cdots + y_k \quad \text{and} \quad N_k = n_1 + \cdots + n_k$$

similarly to the previous Section. Then the max chi-squared statistic is defined by

$$\max \chi^2 = \max(\chi_1^2, \chi_2^2, \ldots, \chi_{K-1}^2)$$

where

$$\chi_k^2 = \frac{N_K^3 (Y_k - Y_K N_k / N_K)^2}{Y_K (N_K - Y_K) N_k (N_K - N_k)} \tag{3.1}$$

Table 3: Accumulated table

Accumulated dose levels	Improved	Not improved	Total
$1 \sim k$	Y_k	$N_k - Y_k$	N_k
$k+1 \sim K$	$Y_K - Y_k$	$N_K - N_k - Y_K + Y_k$	$N_K - N_k$
Total	Y_K	$N_K - Y_K$	N_K

is the usual goodness of fit chi-squared for the kth accumulated Table 3.

Now we are interested in the joint probability

$$F(\chi^2) = \Pr\{\chi_1^2 \le \chi^2, \dots, \chi_{K-1}^2 \le \chi^2\}.$$

Again we have a very efficient and exact algorithm for evaluating F based on the Markov property of the Y_k's. Define a conditional probability

$$F_k(\chi^2|Y_k) = \Pr\{\chi_1^2 \le \chi^2, \dots, \chi_k^2 \le \chi^2|Y_k\}$$

and an initial function

$$F_1(\chi^2|Y_1) = \begin{cases} 1, & \chi_1^2 \le \chi^2, \\ 0, & \text{otherwise.} \end{cases}$$

It should be noted here that χ_k^2 (3.1) is a function of Y_k only given all the marginal totals. Then we have Lemma 2.

Lemma 2.   Recurrence formula 2

$$F_{k+1}(\chi^2|Y_{k+1}) = \begin{cases} \sum_{Y_k} F_k(\chi^2|Y_k) f(Y_k|Y_{k+1}), & \chi_{k+1}^2 \le \chi^2, \\ 0, & \text{otherwise.} \end{cases}$$

where $f(Y_k|Y_{k+1})$, the conditional distribution of Y_k given Y_{k+1}, is the hypergeometric (in the null case) or the extended hypergeometric (in the nonnull case) distribution given all the marginal totals of Table 4.

We have derived the max χ^2 test above based on the goodness of fit chi-squared. It may, however, be more appropriate in this case to consider one-sided test based on

$$\chi_k = \left\{ \left(\frac{1}{N_k} + \frac{1}{N_k^*} \right) \frac{Y_K}{N_K} \left(1 - \frac{Y_K}{N_K} \right) \right\}^{-1/2} \left(\frac{Y_k^*}{N_k^*} - \frac{Y_k}{N_k} \right),$$

Table 4: Subtable for a conditional probability of t_k

Accumulated dose levels	Improved	Not improved	Total
$1 \ldots k$	Y_k	$N_k - Y_k$	N_k
$k+1$	y_{k+1}	$n_{k+1} - y_{k+1}$	n_{k+1}
Total	Y_{k+1}	$N_{k+1} - Y_{k+1}$	N_{k+1}

Table 5: One-sided max χ^2 test applied to Table 2

Dose level	y_i	n_i	χ_k
1	20	36	1.319
2	23	41	2.276 $(=\max \chi_k)$
3	27	36	1.161
4	26	35	-0.043
5	9	14	

Exact one-sided p value for $\max \chi_k = 2.276$ is 0.044.

the square of which is equal to (3.1). This form is more similar to the statistic t_k in (2.1).

Example 2. We apply the method to Table 2 and show the result in Table 5. It suggests that the dose levels 3, 4, 5 will give a higher response rate as compared with the levels 1 and 2. This type of conclusion should be more useful for our interest here as compared with the overall trend tests such as Wilcoxon test.

4. Comparing Treatments Based on the Ordered Categorical Responses

4.1 Max χ^2 Method for a $2 \times K$ Contingency Table

The data in Table 6 are typical in the clinical trial for comparing two treatments. In Table 6 the high categories seem to occur more frequently in the higher dose. To confirm it one may search for a cut point of ordered columns which gives the largest χ^2 value among all the resulting 2×2 tables obtained by cutting and pooling columns at every possible cut point, see Table 7.

From Table 7 the largest χ^2 value is obtained as $\chi^2 = 10.651$ by partitioning Table 6 between the 4th and 5th columns. In the past it

Table 6: Usefulness in a dose-response experiment

Drug	1 Undesirable	2 Slightly undesirable	3 Not useful	4 Slightly useful	5 Useful	6 Excellent
AF3mg	7	4	33	21	10	1
AF6mg	5	6	21	16	23	6

Table 7: χ^2 values for 2×2 tables obtained by cutting and pooling columns

Drug	Accumulated ordered categories					
	(1)	(2 ~ 6)	(1, 2)	(3 ~ 6)	(1 ~ 3)	(4 ~ 6)
AF3mg	7	69	11	65	44	32
AF6mg	5	72	11	66	32	45
Statistic	$\chi^2 = 0.542$		$\chi^2 = 0.001$		$\chi^2 = 3.765$	
Drug	(1 ~ 4)	(5, 6)	(1 ~ 5)	(6)	Total	
AF3mg	65	11	75	1	76	
AF6mg	48	29	71	6	77	
Statistic	$\chi^2 = 10.651$		$\chi^2 = 3.675$			

used to be evaluated as the χ^2 distribution with 1 degree of freedom to give the p value as low as 0.001. It, however, obviously suffers from the false positive problem by choosing the largest value among 5 candidate chi-squared values. Now, in spite of the difference of the basic sampling distribution exactly the same algorithm as given in the previous section can be applied to evaluating the exact p value of the selected maximal chi-squared. It is because we employ the conditional inference given all the marginal totals, see Hirotsu et al. (1992), for example.

For Table 7 it gives 0.0033 as the exact two-sided p value.

4.2 Comparing Ordered Treatments Based on the Ordered Categorical Responses

The data in Table 6 are actually a part of Table 8 which gives again a typical example of data from dose-response experiment.

Table 8 is characterized to have natural ordering in both of rows and columns. The purpose of the experiment is to find out the dose level which is clearly superior to the dose levels below it. Then the

Table 8: Usefulness in a dose-response experiment

Drug	1 Undesirable	2 Slightly undesirable	3 Not useful	4 Slightly useful	5 Useful	6 Excellent
Placebo	3	6	37	9	15	1
AF3	7	4	33	21	10	1
AF6	5	6	21	16	23	6

Table 9: Tables obtained by cutting and pooling rows

| | Ordered Categories | | | | | | |
Accumulated dose levels	1	2	3	4	5	6	Total
(1)	3	6	37	9	15	1	71
(2,3)	12	10	54	37	33	7	153
(1,2)	10	10	70	30	25	2	147
(3)	5	6	21	16	23	6	77
Total	15	16	91	46	48	8	224

max χ^2 method seems to be appropriate for comparing the dose levels. The subtables obtained by cutting and pooling ordered rows are shown in Table 9.

There are several choices of the procedures to deal with the ordered columns. One possibility is to apply max χ^2 method for each of two 2×6 subtables in Table 9. The calculation is shown in Table 10 to give the largest χ^2 value 10.033 among all the possible 2×2 tables. Again an exact algorithm has been obtained for calculating its p value in Hirotsu (1997) and the method is called max max χ^2 since it is searching for the maximal χ^2 in two-ways simultaneously along the rows and columns. The method gives the two-sided p value 0.014 when applied to Table 10.

Another possibility will be to apply a linear rank test such as Wilcoxon test or the cumulative χ^2 test for each of two subtables and use the maximal statistic, which we call here the max Wil and max χ^{*2} tests, respectively. For the cumulative χ^2 test for a two-way contingency table one should refer to Hirotsu (1982, 1983b) and Takeuchi and Hirotsu (1982). When applied to Table 9, those statistics and the corresponding two-sided p values are as given in Table 11, where the notation $W(1, 2; 3)$ implies that the maximal statistic was obtained between the dose levels 2 and 3.

Table 10: Calculating the $\max\max\chi^2$ statistic

Accumulated dose levels	Accumulated ordered categories					
	(1)	(2 ~ 6)	(1, 2)	(3 ~ 6)	(1 ~ 3)	(4 ~ 6)
(1)	3	68	9	62	46	25
(2,3)	12	141	22	131	76	77
	$\chi^2_{11} = 1.01589$		$\chi^2_{12} = 0.11796$		$\chi^2_{13} = 4.46771$	

Accumulated dose levels	(1 ~ 4)	(5, 6)	(1 ~ 5)	(6)	Total
(1)	55	16	70	1	71
(2,3)	113	40	146	7	153
	$\chi^2_{14} = 0.33680$		$\chi^2_{15} = 1.41212$		

Accumulated dose levels	Accumulated ordered categories					
	(1)	(2 ~ 6)	(1, 2)	(3 ~ 6)	(1 ~ 3)	(4 ~ 6)
(1,2)	10	137	20	127	90	57
(3)	5	72	11	66	32	45
	$\chi^2_{21} = 0.00773$		$\chi^2_{22} = 0.01961$		$\chi^2_{23} = 7.88007$	

Accumulated dose levels	(1 ~ 4)	(5, 6)	(1 ~ 5)	(6)	Total
(1,2)	120	27	145	2	147
(3)	48	29	71	6	77
	$\chi^2_{24} = 10.03340$		$\chi^2_{25} = 6.06959$		

Table 11: Multiple comparisons of three dose levels

Test statistic	two-sided p-value
$\max Wil = W(1,2;3) = 2.7629$	0.011
$\max \chi^{*2} = \chi^{*2}(1,2;3) = 24.010$	0.005

The methods introduced here should be called the row-wise multiple comparisons of a two-way table with a natural ordering in rows and another type of row-wise multiple comparisons will be given in Section 6.

5. Max χ^2 Method for Testing the Association Between the Disease and Alleles at Highly Polymorphic Loci

For analyzing the association between the disease and allele frequencies at highly polymorphic loci some statistical tests for a $2 \times J$ contingency table have been proposed. As an example of such a $2 \times J$ table the allele frequencies of the schizophrenia and normal at the

Table 12: Allele frequencies at the D20S95 locus

	97	99	101	103	105	107	109	111	113	115
schizophrenia	4	30	25	10	17	90	34	37	4	1
normal	5	18	28	6	41	67	5	34	5	5

D20S95 locus are given in Table 12.

Among those testing procedures Sham and Curtis (1995) compare by simulation four statistical tests:

1. usual goodness of fit chi-squared,
2. the goodness of fit chi-squared pooling small cell frequencies,
3. multiple comparisons of one cell at a time against others,
4. multiple comparisons of all 2×2 tables made from the original table by pooling columns.

Among them we can give an exact algorithm to evaluate the p value of the third method which we call here the max 1-to-others χ^2. It is obviously appropriate if there is only one susceptibility allele in the locus. Including this all those four tests discussed in Sham and Curtis (1995) are intuitively acceptable. They do not, however, take into account the natural ordering in the number of CA repeats whereas the abnormal extension of CAG repeats has been reported associated with the Huntington's disease. We therefore consider the max χ^2 test proposed in Section 4 for an ordered alternative in a $2 \times K$ contingency table to be appropriate also for this problem. We call the test here as the max accumulated χ^2 test distinguishing it from the max 1-to-others χ^2. We further propose to combine those two tests when the prior knowledge is not sufficient enough to specify one of those two alternative hypotheses (Hirotsu et al., 2001).

The simultaneous analyses of the two closely linked loci in a chromosome have been called a haplotype analysis. Then the frequency data are presented in a $2 \times J \times K$ contingency table as shown in Table 13 but in Sham and Curtis (1995) it has been dealt with as if it were a $2 \times JK$ two-way table.

It is, however, obvious that we need further the analysis of three-way interaction to distinguish the two cases shown in Tables 14(a) and (b). In Table 14(a) the probability model $p_{ijk} = p_{ij.} \times p_{i.k}$ holds suggesting the singularities of the row 2 and column 3 associated with the disease without interaction effects of the row and column on the

Table 13: Haplotype allele frequencies at loci DXS548 and FRAXAC2 in fragile X and normal chromosomes Fragile X ($i = 1$)

j \ k	1	2	3	4	5	6	7	8	9	10	total
1	0	0	0	0	0	0	0	0	0	0	0
2	1	3	0	0	9	16	12	0	0	0	41
3	0	5	9	1	11	1	3	0	0	0	30
4	0	0	0	0	0	0	1	0	0	0	1
5	0	0	0	0	0	0	1	0	0	0	1
6	0	1	0	1	9	3	14	0	0	0	28
7	0	0	0	0	1	0	0	0	0	0	1
total	1	9	9	2	30	20	31	0	0	0	102

Normal ($i = 2$)

j \ k	1	2	3	4	5	6	7	8	9	10	total
1	0	0	0	0	0	2	0	0	0	0	2
2	1	7	5	1	17	67	7	4	1	0	110
3	0	4	6	0	3	8	1	0	0	0	22
4	0	1	0	0	0	0	0	0	0	0	1
5	0	0	0	0	0	1	1	0	0	0	2
6	0	2	0	0	3	6	2	0	0	1	14
7	0	0	0	0	0	0	2	0	0	0	2
total	1	14	11	1	23	84	13	4	1	1	153

disease. In other words the row and the column are conditionally independent given $i = 1$(normal) or 2(disease). On the other hand Table 14(b) suggests the interaction effect pointing out the singularity of the $(2,3)$ cell associated with disease and in this case the separate analyses of marginal tables collapsing rows or columns are quite misleading. We therefore propose a new approach extending the max 1-to-others χ^2 and the max accumulated χ^2 to the analysis of a three-way interaction in the $2 \times J \times K$ table. As a by-product a factorization theory of the probability distribution of three way cell frequencies is obtained, see Hirotsu et al. (2001) for details.

Some applications have been presented at the Symposium.

Table 14: Configuration of p_{ijk} (normalizing constant omitted)

(a)

Normal($i = 1$)						Disease($i = 2$)					
locus 1	locus 2 (k)					locus 1	locus 2 (k)				
(j)	1	2	3	4	5	(j)	1	2	3	4	5
1	1	1	1	1	1	1	1	1	2	1	1
2	1	1	1	1	1	2	2	2	4	2	2
3	1	1	1	1	1	3	1	1	2	1	1
4	1	1	1	1	1	4	1	1	2	1	1

(b)

Normal($i = 1$)						Disease($i = 2$)					
locus 1	locus 2 (k)					locus 1	locus 2 (k)				
(j)	1	2	3	4	5	(j)	1	2	3	4	5
1	1	1	1	1	1	1	1	1	1	1	1
2	1	1	1	1	1	2	1	1	4	1	1
3	1	1	1	1	1	3	1	1	1	1	1
4	1	1	1	1	1	4	1	1	1	1	1

6. Comparing Treatments Based on Repeated Measures

The data of Table 15 are the total cholesterol amount measured on 23 subjects at every 4 weeks for 6 periods. Let the repeated measurements for the ith subject assigned to the hth treatment be denoted generally by $y_{hi} = (y_{hi1}, \ldots, y_{hip})'$, $i = 1, \ldots, n_h$; $h = 1, \ldots, t$. Note that $t = 2$ and $p = 6$ in the example of Table 15. In comparing the treatments based on these data, applying t tests repeatedly to the data at different periods of course suffer from the multiplicity problem similar to the repeated χ^2 tests for an ordered $2 \times K$ contingency table. There might also be difficulties in combining the different results at different periods into a single consistent conclusion.

The conventional approach assumes a multivariate model

$$y_{hi} = \mu_h + \varepsilon_{hi} \tag{6.1}$$

and compares the mean vectors $\mu_1, \ldots, \mu_t$ assuming that ε_{hi} are independently and identically distributed as $N(0, \Omega)$ with the serial correlations within the subjects. To reduce the number of unknown parameters, a simple model such as an AR model is sometimes assumed for the serial correlation, see Ware (1985), for example. If the covariance matrix Ω reduces to $\Omega = [\sigma_0^2 I + \sigma_1^2 jj']$, I being an identity matrix and j a vector of 1's, then the standard analysis for a split plot design can be applied with the treatment and period as the first and the second order factors, respectively, and the subject as block (Aitkin, 1981 and Wallenstein, 1982). This would, however, be rarely the case since serial measurements conceptually contradict with the assumption of randomization over plots and make the permutation invariant procedures unattractive.

The general two stage model by Laird and Ware (1982) permits a more general description of covariance structures through the distribution of subject profiles and should be an efficient approach if from past experience there is available a reasonable regression model to describe the subject profiles. In its simplest form, for example, a simple linear regression model

$$y_{ij} = \alpha + \beta x_j + \varepsilon_{ij}$$

is assumed for the jth measurement on the ith individual. Then supposing that there is some variation in the slope between individuals, β is replaced by $\beta + b_i$ where β is the population mean slope and b_i is the individual departure from the mean slope. Then we have $E(y_{ij}) = \alpha + \beta x_j$ and the covariance structure

$$\mathrm{Cov}(y_{ij}.y_{ij'}) = \sigma_b^2 x_j x_{j'} + \sigma^2 \delta_{jj'},$$

where $\sigma_b^2 = V(b_i), \sigma^2 = V(\varepsilon_{ij})$ and $\delta_{jj'} = 1$ if $j = j'$ and $= 0$ otherwise, see Crowder and Hand (1990, Chapter 6) for additional details.

There are, however, some cases where the prior information is too vague to assume a definite polynomial model in time x but it is still reasonable to assume some systematic trend along the time axis. In these cases, a reasonable approach will be to assume a simple mixed effects model

$$y_{hi} = \mu_{hi} + \varepsilon_{hi} \qquad (6.2)$$

with the μ_{hi} independently distributed as $N(\mu_h, \Omega_h)$, and we can try to incorporate the natural ordering along the time axis into the analysis. It should be noted that now the μ_{hi} has suffix i to distinguish the individual effects. The pure measurement errors at different periods are reasonably assumed to be independent of each other and also mutually independent of the μ_{hi}. Therefore, the ε_{hi} are assumed to be independently distributed as $N(0, \sigma^2 I)$. Given h, this is just Schffé's (1959,

Chapter 8) two-way mixed effects model with subject as a random factor, period as a fixed factor and without replication. In this case, as in the two stage model, the covariances among repeated measurements are thought to arise from the inhomogeneity of the individual profiles along the time axis rather than the serial correlations.

Now if there is no treatment effect, then the expected profile of each subject should be stable over the period with some random fluctuations around it. On the other hand if a treatment has any effect, individual profiles should change. The effect may, however, be not homogeneous over the subjects, who might therefore be classified into several groups, improved, invariant, deteriorated and so on. This means that the treatment should affect the covariance structure of the μ_{hi} as well as its mean vector. Thus for evaluating treatment effects it can be quite misleading to compare mean vectors assuming the equality of the covariance structures. This is the conceptual difference between model (6.2) and the naive multivariate normal model (6.1). The equality of the covariances should of course be tested preliminarily in the usual approach, see Morrison (1976, Chapter 4), for example, but in the latter case it shall be the main objective of the experiment. It should also be noted that for comparing covariance matrices, the usual permutation invariant tests are inappropriate since here we are concerned with some systematic change in the profiles along the time axis.

A possible approach is to consider the subjects as a variation factor and classify them into homogeneous subgroups based on their time profiles. This is done by the row-wise multiple comparisons for interaction in the two-way table with the subject as row factor, period as ordered column categories and without replication. The classification naturally induces an ordering among the subgroups and each treatment is then characterized by distribution of its own subjects over those subgroups. This gives an interesting example of the multiple comparisons methods for interaction effects assuming some systematic order effects only in columns whereas we assumed natural ordering in both of rows and columns in Section 4.2.

Now, renumbering the subjects throughout the treatments, we denote by y_{kj} the measurement at the jth period of the kth subjects, $k = 1, \ldots, m$, $m = \sum n_h$. The basic statistic to measure the difference of the profiles between subjects k and k' is given by

$$
\begin{aligned}
S(k;k') &= \| (\boldsymbol{k}(k;k')' \otimes C^{*\prime})\boldsymbol{y} \|^2 \\
&= \frac{1}{2} \sum_{l=1}^{p-1} \frac{pl}{p-l} \left\{ \frac{1}{l} \sum_{j=1}^{l} (y_{kj} - y_{k'j}) - (\bar{y}_{k.} - \bar{y}_{k.'}) \right\}^2,
\end{aligned}
\tag{6.3}
$$

where $\boldsymbol{y}$ is the vector of y_{kj} arranged in the dictionary order, $\boldsymbol{k}(k;k') = 2^{-1/2}(0 \cdots 0 \; -1 \; 0 \cdots 0 \; 1 \; 0 \cdots 0)$ a vector to define a contrast between the subject k and k', and $C^{*\prime}$ a matrix obtained by normalizing the rows of $(D_p' D_p)^{-1} D_p'$ defined in Section 2. In (6.3) $C^{*\prime}$ is introduced for detecting the systematic departure along time axis. This is the partitioned sum of squares of the cumulative chi-squared statistic

$$
\begin{aligned}
S_{R \times C}^* &= \| (R' \otimes C^{*\prime})\boldsymbol{y} \|^2 \\
&= \sum_{k=1}^{m} \sum_{l=1}^{p-1} \frac{pl}{p-l} \left\{ \frac{1}{l} \sum_{j=1}^{l} (y_{kj} - \bar{y}_{.j}) - (\bar{y}_{k.} - \bar{y}_{..}) \right\}^2
\end{aligned}
$$

introduced in Hirotsu (1978) for interaction effects between rows and columns to be explained by the systematic departure in profiles of subjects k and k'. The sum of squares $S(k;k')$ between the subjects k and k' for Table 15 are given in Table 16 where subjects are rearranged so that those with smaller between sum of squares are located closer.

The sum of squares between the two subgroups of subjects, $I_1 = (1, \ldots, q_1)$ and $I_2 = (q_1+1, \ldots, q_1+q_2)$, say, is given in vector notation as

$$
S(I_1;I_2) = \| (\boldsymbol{k}'(I_1;I_2) \otimes C^{*\prime})\boldsymbol{y} \|^2,
\tag{6.4}
$$

where the notation $\boldsymbol{k}'(I_1;I_2)$ will be self-explanatory.

For every choice of subgroups I_1 and I_2 the between groups sum of squares $S(I_1;I_2)$ is bounded above by the maximum root of the Wishart matrix $W(\sigma^2 C^{*\prime} C^*, m-1)$ under the null hypothesis of homogeneity of subjects (Hirotsu, 1983a). Therefore, Scheffé type multiple comparisons can be performed by referring to the Wishart distribution after canceling out the unknown σ^2. This is done by dividing $S(I_1;I_2)$ by some appropriate estimator of σ^2. It is important that the denominator sum of squares is to be less affected by a systematic departure among subjects.

Table 15: Total cholesterol amounts

Treatment	Subject	Period 1	2	3	4	5	6
Drug	1	317	280	275	270	274	266
	2	186	189	190	135	197	205
	3	377	395	368	334	338	334
	4	229	258	282	272	264	265
	5	276	310	306	309	300	264
	6	272	250	250	255	228	250
	7	219	210	236	239	242	221
	8	260	245	264	268	317	314
	9	284	256	241	242	243	241
	10	365	304	294	287	311	302
	11	298	321	341	342	357	335
	12	274	245	262	263	235	246
Placebo	13	232	205	244	197	218	233
	14	367	354	358	333	338	355
	15	253	256	247	228	237	235
	16	230	218	245	215	230	207
	17	190	188	212	201	169	179
	18	290	263	291	312	299	279
	19	337	337	383	318	361	341
	20	283	279	277	264	269	271
	21	325	257	288	326	293	275
	22	266	258	253	284	245	263
	23	338	343	307	274	262	309

Table 16: Between sum of squares $S(k;k')$

$k\backslash k'$	8	11	4	18	7	19	2	22	21	20	17	16	15	14	13	12	6	10	9	1	3	23
23	16.7	11.7	8.8	7.7	7	5.9	4.3	4.6	4.9	2.9	3.6	3.7	2.1	2.1	4.4	2.0	1.5	1.2	0.9	1.0	0.7	
3	18.4	11.5	8.5	7.4	6.6	5.7	5.5	4.8	4.6	3.1	3.0	3.2	2.1	2.7	5.1	1.8	1.7	1.8	1.1	1.0		
1	12.7	8.1	6.0	4.2	3.9	3.5	3	2.3	1.8	1.3	1.7	1.6	0.8	1.0	2.5	0.5	0.4	0.3	0.03			
9	11.9	7.6	5.6	3.8	3.6	3.2	2.6	2.0	1.7	1	1.5	1.4	0.6	0.7	2.2	0.07	0.3	0.4				
10	13.7	10.1	8.1	5.6	5.4	5.0	3.6	3.5	2.6	2.2	3.2	2.8	1.7	1.6	3.3	0.4	1.1					
6	9.7	5.5	3.7	2.6	2.2	2.1	1.9	1.0	1.1	0.5	0.7	0.8	0.3	0.3	1.3	1.4						
12	9.7	5.2	3.4	2.2	1.9	1.9	2.2	1.0	0.9	0.5	0.4	0.6	0.4	0.5	1.3							
13	2.3	4.3	1.5	1.8	0.8	0.5	0.3	0.5	1.4	0.4	1.0	0.7	0.8	0.4								
14	7.1	4.0	2.7	1.9	1.5	1.1	0.7	0.6	1.2	0.1	0.7	0.5	0.1									
15	8.0	4.1	2.7	1.0	1.5	1.1	1.1	0.7	1.2	0.1	0.5	0.3										
16	6.8	2.8	1.6	1.2	0.7	0.5	1.3	0.5	0.8	0.2	0.2											
17	7.6	3.0	1.5	1.2	0.8	0.8	1.8	0.5	1.0	0.4												
20	6.4	3.1	2.0	0.9	1.0	0.7	0.7	0.3	0.9													
21	7.0	3.6	1.2	0.5	1.1	1.8	2.5	0.7														
22	5.0	2.0	1.8	1.7	0.4	0.7	1.1															
5	9.1	3.2	2.3	1.9	1.2	1.3	3.2															
2	4.4	3.0	0.6		1.6	0.9																
19	4.3	1.3			0.4																	
7	3.8	0.8	0.4	0.09																		
18	3.5	0.9	0.7																			
4	3.5	0.4																				
11	2.0																					
8																						

As one of those,

$$\chi^{-2} = \sum_{k=1}^{m} \sum_{j=1}^{p-1} \left\{ \frac{y_{kj-1} - y_{kj} - (\bar{y}_{.j-1} - \bar{y}_{.j})}{\rho_j} \right\}^2$$

is introduced in Hirotsu (1991), where $\rho_j = [p/\{j(p-j)\}]^{\frac{1}{2}}$ is the normalizing constant. χ^{-2} possesses inverse characteristic to the cumulative chi-squared statistic χ^{*2} introduced in Hirotsu (1978) as shown by the expansion,

$$\chi^{-2} = \left\{ \frac{1 \times 2}{p} \chi_{(1)}^2 + \frac{2 \times 3}{p} \chi_{(2)}^2 + \cdots + \frac{(p-1) \times p}{p} \chi_{(p-1)}^2 \right\},$$

where $\chi_{(j)}^2$ is the chi-squared component for the Chebyshev jth orthogonal polynomial each with $m - 1$ degrees of freedom in this case. This suggests that the χ^{-2} would be appropriate for evaluating σ^2 when some systematic departure like linear trend exist among subjects but with some loss of degrees of freedom in the null case, see also Hirotsu (1986). On the other hand it should be noted that the cumulative chi-squared statistic χ^{*2} is expanded in

$$\chi^{*2} = p \left\{ \frac{1}{1 \times 2} \chi_{(1)}^2 + \frac{1}{2 \times 3} \chi_{(2)}^2 + \cdots + \frac{1}{(p-1)p} \chi_{(p-1)}^2 \right\}.$$

Since asymptotically the maximum root of $W(\sigma^2 C^{*'} C^*, m-1)$ is distributed as $(p/2)\chi_{(1)}^2$, $S(I_1; I_2)/\chi^{-2}$ is bounded above by

$$\frac{p}{2}\chi_{(1)}^2 \Big/ \left\{ \frac{1 \times 2}{p} \chi_{(1)}^2 + \frac{2 \times 3}{p} \chi_{(2)}^2 + \cdots + \frac{(p-1) \times p}{p} \chi_{(p-1)}^2 \right\}. \quad (6.5)$$

A method is given in Hirotsu (1991) to classify subjects into homogeneous subgroups and also to evaluate the significance of the classification referring to the null distribution (6.5). Each of the obtained classes is characterized by its own response pattern. The response pattern Γ_{Ij} of a class I of size q_I is estimated by the block interaction effects

$$\hat{\Gamma}_{Ij} = \frac{1}{q_I} \sum_{i \in I} (y_{ij} - \bar{y}_{i.} - \bar{y}_{.j} + \bar{y}_{..}), \quad j = 1, \ldots, p. \quad (6.6)$$

Usually there arises a natural ordering in those classes induced by the original natural ordering along the time axis and also by the systematic nature of the distance statistics (6.3) and (6.4).

Table 17: Observed distributions for the drug and the placebo

	Class		
Treatment	I_1:Improved	I_2:Invariant	I_3:Deteriorated
Drug	4	4	4
Placebo	1	9	1

For the data of Table 15, the classification into $I_1 = (1, 3, 9, 10, 23)$, $I_2 = (2, 5, 6, 12, 13, 14, 15, 16, 17, 19, 20, 21, 22)$ and $I_3 = (4, 7, 8, 11, 18)$ is obtained at significance level 0.05. The interactions $(y_{ij} - \bar{y}_{k.} - \bar{y}_{.j} + \bar{\bar{y}}_{..})$ are plotted in Figure 1 for each of the three classes I_1, I_2 and I_3. The estimated response pattern for each of the three classes are calculated by (6.6) and plotted also in Fig. 1 with thick lines. Obviously I_1 is the improved group, I_3 the deteriorated group and I_2 is invariant.

The observed distribution of subjects from each of the two treatment groups over these three classes is given in Table 17. Pearson's chi-squared value 5.49 for Table 17 is not significant at 0.05 level when compared to the chi-squared distribution with two degrees of freedom. It is, however, purely the quadratic component along the column categories suggesting that the difference between the active drug and the placebo is not in the mean profiles but in the dispersion around the mean profiles. Therefore if we had employed the multivariate normal model to compare the mean vectors assuming the equality of the covariance matrices, we would have failed to prove the difference. On the other hand, both the modified likelihood ratio test (Bartlett, 1937; Anderson, 1984) and Nagao's (1973) invariant test for comparing two Wishart matrices have significance levels approximately 0.25 and are not successful in detecting such a systematic difference in covariance matrices along the time axis.

For detailed explanation of the method and for other examples see Hirotsu (1991). Analysis of residuals are also recommended there to verify the homogeneity within the subgroups by some nonsystematic statistic, since the classification is based on the systematic statistic less affected by the short term deviations. The two-step procedure of

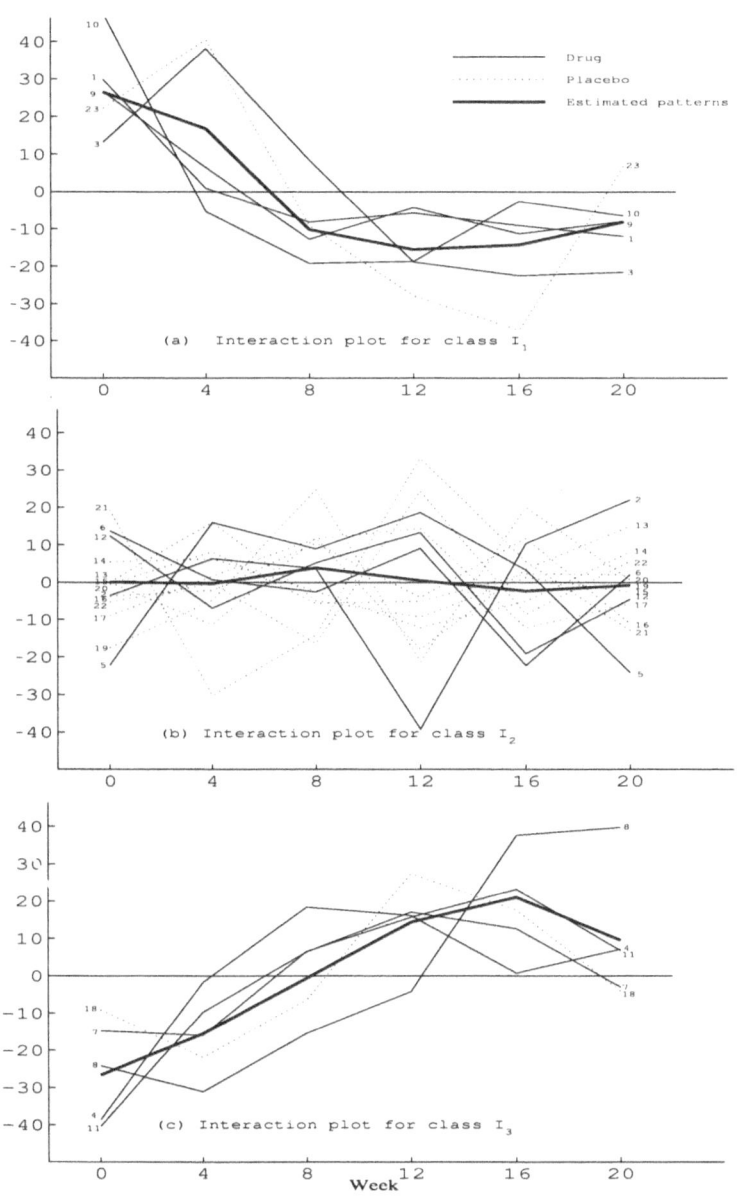

Figure 1:The interaction plots and the estimated response patters

modelling, first analyzing the systematic effects by some systematic statistic less affected by the short term deviations and then analyzing residuals by some nonsystematic statistic to reveal outliers, seems to be useful generally for modelling an ordered data set not restricted to the particular problem here, see Hirotsu (1990) for other examples.

7. Concluding Remarks

Varieties of the multiplicity problems and some statistical approaches in the analysis of clinical trials have been introduced. In particular $\max t$ and $\max \chi^2$ type test statistics are very useful for comparing ordered treatments and for testing the shape constraints for a dose-response curve such as monotone, convexity or sigmoidicity. Those shape constraints are in turn useful for forming the simultaneous confidence intervals for the responses (Hirotsu and Srivastava, 2000). The statistics are also useful for comparing treatments based on the ordered categorical responses or repeated measurements where some systematic change is expected along the time axis such as monotonic or convex trend. It is regarded as an approach to the analysis of interaction effects based on the $\max t$ and $\max \chi^2$. An extension of the $\max \chi^2$ test to the analysis of three-way interaction in a three-way contingency table with ordered categories and the related distribution theory seem to be interesting. It seems, however, that there are still a very few papers dealing with the ordered hypotheses on the three-way interaction.

References

Aitkin, M. (1981). Regression models for repeated measurements (Response to Query). *Biometrics*, **37**, 831-832.

Anderson, T. W. (1984). *An introduction to multivariate statistical analysis, 2nd edition*. New York: Wiley.

Armitage, P. and Parmar, M. (1986). Some approaches to the problem of multiplicity in a clinical trial. *Proceedings, Invited Papers, 13th International Biometric Conference*.

Bartlett, M. S. (1937). Properties of sufficiency and statistical tests. *Proceedings Royal, Society, London*, **A 160**, 268-282.

Crowder, M. J. and Hand, D. J. (1990). *Analysis of repeated measures*. London: Chapman and Hall.

Dunnett, C. W. (1964). Comparing several treatments with a control. *Biometrics*, **20**, 482-491.

Hawkins, A. J. (1977). Testing a sequence of observations for a shift in location. *Journal of the American Statistical Association*, **72**, 180-186.

Hirotsu, C. (1978). Ordered alternatives for interaction effects. *Biometrika*, **65**, 561-570.

Hirotsu, C. (1982). Use of cumulative efficient scores for testing ordered alternatives in discrete models. *Biometrika*, **69**, 567-577.

Hirotsu, C. (1983a). An approach to defining the pattern of interaction effects in a two-way layout. *Annals of Institute of Statistical Mathematics*, **A35**, 77-90.

Hirotsu, C. (1983b). Defining the pattern of association in two-way contingency tables. *Biometrika*, **70**, 579-589.

Hirotsu, C. (1986). Cumulative chi-squared statistic as a tool for testing goodness of fit. *Biometrika*, **73**, 165-173.

Hirotsu, C. (1990). Discussion on "A critical look at accumulation analysis and related methods" by Hamada, M. and Wu, C. F. J. *Technometrics*, **32**, 133-136.

Hirotsu, C. (1991). An approach to comparing treatments based on repeated measures. *Biometrika*, **78**, 583-594.

Hirotsu, C. (1993). Beyond analysis of variance techniques : Some applications in clinical trials. *International Statistical Review*, **61**, 183-201.

Hirotsu, C. (1997). Two-way change point model and its application. *Australian Journal of Statistics*, **39**, 205-218.

Hirotsu, C., Kuriki, S. and Hayter, A. J. (1992). Multiple comparison procedures based on the maximal component of the cumulative chi-squared statistic. *Biometrika*, **79**, 381-392.

Hirotsu, C. and Srivastava, M. S. (2000). Simultaneous confidence intervals based on one-sided max t test. *Statistics and Probability Letters*, **49**, 25-37.

Hirotsu, C., Aoki, S., Inada, T. and Kitao, T. (2001). An exact test for the association between the disease and alleles at highly polymorphic loci - with particular interest in the haplotype analysis. *Biometrics*, **57**, 148-157.

Hirotsu, C. and Marumo, K. (2002). Changepoint analysis as a method for isotonic inference. To appear in *the Scandinavian Journal of Statistics*.

Laird, N. M. and Ware, J. H. (1982). Random-effects model for longitudinal data. *Biometrics*, **38**, 963-974.

Lewis, J. A. (1999). Statistical principles for clinical trials (ICH E9): an introductory note on an international guideline. *Statistics in Medicine*, **18**, 1903-1942.

Marcus, R. (1976). The powers of some tests of the equality of normal means against an ordered alternative. *Biometrika*, **63**, 177-183.

Marcus, R., Peritz, E., and Gabriel, K. R. (1976). On closed testing procedure with special reference to ordered analysis of variance. *Biometrika*, **63**, 655-660.

Morrison, D. F. (1976). *Multivariate statistical methods, 2nd edition.* New York: McGraw-Hill.

Nagao, H. (1973). On some test criteria for covariance matrix. *Annals of Statistics*, **1**, 700-709.

Scheffé, H. (1959). *The analysis of variance.* New York: Wiley.

Sham, P. C. and Curtis, D. (1995). Monte Carlo tests for associations between disease and alleles at highly polymorphic loci. *Annals Human Genetics*, **59**, 97-105.

Takeuchi, K. and Hirotsu, C. (1982). The cumulative chi-squares method against ordered alternatives in two-way contingency tables. *Reports of Statistical Application Research, Japanese Union of Scientists and Engineers*, **29**, 3, 1-13.

Wallenstein, S. (1982). Regression models for repeated measurements (Reader reaction). *Biometrics*, **38**, 849-850.

Ware, J. H. (1985). Linear models for the analysis of longitudinal studies. *American Statisticians*, **39**, 95-101.

Williams, D. A. (1971). A test for differences between treatment means when several dose levels are compared with a zero dose control. *Biometrics*, **27**, 103-117.

A Probability Analysis of the Playoff System
in Sumo Tournaments

Anthony J. Hayter

School of Industrial and Systems Engineering
Georgia Institute of Technology
Atlanta, Georgia 30332-0205
U.S.A.

Summary: If three rikishi are tied at the conclusion of a sumo tournament, then a playoff system is implemented to determine who will be the tournament champion. This system has the interesting property that the two rikishi who contest the first fight have an advantage over the third rikishi. Some probability calculations are presented which analyse this playoff system and which assess the relative advantages and disadvantages of the starting positions. The number of fights needed to determine the champion is also considered, and comparisons are made with a fairer playoff protocol. Extensions are also considered to situations with more than three rikishi.

1. Introduction

Consider the problem of determining a single winner out of three contestants when the contestants may fight each other only in pairs. That is to say that a series of fights can be scheduled between any two of the contestants at a time, with each fight resulting in one of the contestants being the winner and the other being the loser. In general terms, the objective is to determine a tournament winner in a fair manner and in the shortest possible number of fights.

The simplest protocol would be to have two contestants fight each other, A and B say, with the winner then fighting the third contestant C. The winner of the second fight would be declared the tournament winner. This protocol produces a result after only two fights, but obviously it is not a fair protocol since contestant C has an advantage compared with the other two contestants. Therefore in practice more complicated protocols should be considered.

This problem arises at the end of a sumo tournament when three

27

contestants (rikishi) are tied with the best records. Subsequent playoff fights are then scheduled to determine the tournament winner, and for practical purposes there is clearly an advantage to being able to declare a winner after as few fights as possible. The official protocol for the sumo playoff system for three rikishi can be described as follows.

Sumo Protocol

The three rikishi randomly draw one of three papers which have marks Higashi (East), Nishi (West) or a circle. The East and West rikishi then fight each other and the winner of that match then fights the rikishi who sat out the first match. This process continues with the winner of contest i remaining to fight contest $i + 1$ against the rikishi who sat out fight i. As soon as one rikishi wins two consecutive fights then that rikishi is declared the tournament winner and the playoff is finished.

To understand this protocol suppose that the first two contestants are A and B and that contestant A wins. The second fight will then be between A and C. If A wins the second fight then the playoff is finished and A is the tournament winner. However, if C wins the second fight then a third fight is scheduled between C and B, and if C wins this third fight then the playoff is finished and C is the tournament winner. On the other hand, if B wins the third fight then an interesting position has arisen since each contestant has won once and lost once. A fourth fight will then be scheduled between B and A, and according to the protocol, if B wins then the playoff is finished and B is the tournament winner.

In practice the playoff is often decided after just two fights. However, an interesting playoff occurred in March 1990 when rikishi Hokuto-umi lost the first fight against rikishi Kirishima. Unfortunately for Kirishima, though, he lost the second fight to rikishi Konishiki. Hokuto-umi was then able to win the third fight against Konishiki, and the fourth fight against Kirishima, again, to be declared the tournament champion.

This protocol is interesting because as was noted, if three fights have been completed and the tournament winner has not yet been decided, then the three contestants have equal records. However, the tournament can be decided after only one more fight. Intuitively, it can be seen that this protocol provides an advantage for the two contestants who are in the first fight compared with the contestant who must sit out the first fight. Of course, since the starting positions are selected at random it may be considered to be fair in any case, but it is also desirable that a good protocol should be fair conditional on the starting positions.

A fair protocol would require that the probabilities of winning the tournament are not dependent on the starting positions. For this to be achieved, if three fights have been completed and the tournament winner has not yet been decided with the three contestants having equal records, the subsequent protocol should be equivalent to starting all over again. Specifically, a tournament winner cannot be determined after only one more fight.

There has been some discussion in the statistical literature relating to tournaments between three players which has some relevance to the sumo playoff system. In a ranking and selection setting, Sobel and Weiss (1970) discuss this "drop the loser" scheduling rule for tournaments with three players, and further analysis is provided by Bechhofer (1970). Brace and Brett (1975) refer to decision procedures of this kind as "king of the castle" procedures and discuss various types of them.

In section 2 some probability calculations are made for the sumo protocol which investigate the relative advantages and disadvantages of the starting positions and the number of fights required to determine a tournament winner. Comparisons are also made with a fairer protocol. In section 3 an extension of the sumo protocol to situations with more than three contestants is considered.

2. A Probability Analysis of the Sumo Protocol

Suppose that the three contestants are labelled A, B and C, and that each time contestants A and B fight, say, there is a constant probability p_{AB} that A beats B. The outcomes of the fights then depend upon the six probabilities p_{AB}, $p_{BA} = 1 - p_{AB}$, p_{AC}, $p_{CA} = 1 - p_{AC}$, p_{BC} and $p_{CB} = 1 - p_{BC}$. The subsequent analysis assumes that these probabilities remain constant, so that factors such as the fatigue or the morale of the contestants are not considered.

2.1 General Results

For $i, j, k \in \{A, B, C\}$ with $j \neq k$, let ρ_{jk}^i be defined to be the probability that contestant i wins the tournament conditional on contestant j having just beaten contestant k, and the tournament winner having not yet been decided. In other words, these probabilities of tournament victory represent the situation when contestant j has just beaten contestant k and is waiting to fight against the other contestant. Clearly for each $j, k \in \{A, B, C\}$ with $j \neq k$, $\rho_{jk}^A + \rho_{jk}^B + \rho_{jk}^C = 1$, and these probabilities are given in Theorem 1.

Theorem 1

$$\rho_{AB}^A = \frac{p_{AC}}{1 - p_{CA}p_{BC}p_{AB}},$$

$$\rho_{AB}^B = \frac{p_{CA}p_{BC}p_{BA}}{1 - p_{CA}p_{BC}p_{AB}},$$

$$\rho_{AB}^C = \frac{p_{CA}p_{CB}}{1 - p_{CA}p_{BC}p_{AB}}.$$

Proof of Theorem 1

Notice that by considering the outcome of the subsequent fight between A and C it can be seen that

$$\rho_{AB}^A = p_{AC} + p_{CA}\rho_{CA}^A,$$

and similarly the expressions

$$\rho_{CA}^A = p_{BC}\rho_{BC}^A$$

and

$$\rho_{BC}^A = p_{AB}\rho_{AB}^A$$

can be derived. These equations can be solved to provide the values of ρ_{AB}^A, ρ_{CA}^A and ρ_{BC}^A as given in the theorem (subject to some permutations of the letters A, B and C). This completes the proof of Theorem 1.

Next for $i, j, k \in \{A, B, C\}$ with $j \neq k$ define $\tau_{jk}^i = \tau_{kj}^i$ to be the probability that contestant i wins the tournament conditional on the first fight being between contestants j and k. Again for each $j, k \in \{A, B, C\}$ with $j \neq k$, $\tau_{jk}^A + \tau_{jk}^B + \tau_{jk}^C = 1$, and these probabilities are given in Theorem 2.

Theorem 2

$$\tau_{AB}^A = \frac{p_{AB}p_{AC}}{1 - p_{AB}p_{BC}p_{CA}} + \frac{p_{AB}p_{AC}p_{BA}p_{CB}}{1 - p_{BA}p_{CB}p_{AC}},$$

$$\tau_{AB}^B = \frac{p_{AB}p_{BC}p_{BA}p_{CA}}{1 - p_{AB}p_{BC}p_{CA}} + \frac{p_{BC}p_{BA}}{1 - p_{BA}p_{CB}p_{AC}},$$

$$\tau_{AB}^{C} = \frac{p_{AB}p_{CA}p_{CB}}{1 - p_{AB}p_{BC}p_{CA}} + \frac{p_{BA}p_{CA}p_{CB}}{1 - p_{BA}p_{CB}p_{AC}}.$$

Proof of Theorem 2

Notice that by considering the outcome of the first fight it can be seen that

$$\tau_{AB}^{A} = p_{AB}\rho_{AB}^{A} + p_{BA}\rho_{BA}^{A},$$

and similarly

$$\tau_{BC}^{A} = p_{BC}\rho_{BC}^{A} + p_{CB}\rho_{CB}^{A}.$$

Using the values of the ρ_{jk}^{i} provided in Theorem 1 these equations provide the required values of the τ_{jk}^{i}. This completes the proof of Theorem 2.

It is interesting to examine how the tournament winning probabilities τ_{jk}^{i} depend upon the fight outcome probabilities p_{ij} and the choice of the two contestants in the first fight. Theorem 3 shows that it is always best for a contestant to participate in the first fight, regardless of the actual values of the fight outcome probabilities p_{ij}.

Theorem 3

The inequality $\tau_{AB}^{A} \geq \tau_{BC}^{A}$ holds for all values of the p_{ij}. The inequality is strict if all of the p_{ij} are non-zero.

Proof of Theorem 3

Using the expressions provided in Theorem 2 it can be shown that

$$\tau_{AB}^{A} - \tau_{BC}^{A} = \frac{p_{AB}p_{AC}p_{CB}(p_{BA}(1 - p_{AC}p_{CB}) + p_{AB}^{2}p_{CA}p_{BC})}{(1 - p_{AB}p_{BC}p_{CA})(1 - p_{BA}p_{CB}p_{AC})} \geq 0,$$

and it can be seen that the inequality is strict if all of the p_{ij} are non-zero. This completes the proof of Theorem 3.

The positive differences $\tau_{AB}^{A} - \tau_{BC}^{A}$ and $\tau_{AC}^{A} - \tau_{BC}^{A}$ measure the amount of advantage that contestant A gains by being scheduled in the first fight (against contestant B or against contestant C respectively) rather than having to sit out the first fight. These will be illustrated with some examples later on.

From the point of view of contestant A, the probabilities p_{AB} and p_{AC} measure the strength that the contestant has against the other two contestants. However the probability p_{BC}, that is the relative strengths of contestants B and C when they fight each other, also has an effect on the chance of contestant A winning the tournament. For fixed values of p_{AB} and p_{AC}, Theorems 4 and 5 examine how the probability p_{BC} effects the chance of contestant A winning the tournament. Theorem 4 addresses the case when contestant A participates in the first fight, and Theorem 5 addresses the case when contestant A sits out the first fight.

Theorem 4

For fixed values of p_{AB} and p_{AC}, the value of τ_{AB}^A is maximised when either $p_{BC} = 0$ or $p_{BC} = 1$.

Proof of Theorem 4

Using the expression given in Theorem 2 and substituting $p_{CB} = 1 - p_{BC}$ it can be shown that

$$\frac{d\tau_{AB}^A}{dp_{BC}} = \frac{p_{AB}^2 p_{AC} p_{CA}}{(1 - p_{AB} p_{BC} p_{CA})^2} - \frac{p_{AB} p_{AC} p_{BA}}{(1 - (1 - p_{BC}) p_{BA} p_{AC})^2},$$

which can be seen to be an increasing function of p_{BC}. Consequently,

$$\frac{d^2 \tau_{AB}^A}{dp_{BC}^2} \geq 0$$

which is sufficent to prove the theorem. This completes the proof of Theorem 4.

When $p_{BC} = 0$ then

$$\tau_{AB}^A = \frac{p_{AB} p_{AC} (1 + p_{BA} p_{CA})}{1 - p_{BA} p_{AC}},$$

and when $p_{BC} = 1$ then

$$\tau_{AB}^A = \frac{p_{AB} p_{AC}}{1 - p_{AB} p_{CA}},$$

so that according to Theorem 4 the maximum value of τ_{AB}^A will be the larger of these two values. The minimum value of τ_{AB}^A is not so easy to characterise, and may be at one of these extreme values of p_{BC}, or may occur for some p_{BC} strictly betwen zero and one.

Theorem 5

For fixed values of p_{AB} and p_{AC}, the value of τ_{BC}^A is maximised when either $p_{BC} = 0$ or $p_{BC} = 1$, and is minimised when

$$p_{BC} = \frac{p_{BA} p_{AC}}{p_{CA} p_{AB} + p_{BA} p_{AC}}.$$

Proof of Theorem 5

In this case it can be shown that

$$\frac{d\tau^A_{BC}}{dp_{BC}} = p_{AB}p_{AC}\left(\frac{1}{(1-p_{AB}p_{BC}p_{CA})^2} - \frac{1}{(1-(1-p_{BC})p_{BA}p_{AC})^2}\right),$$

which again can be seen to be an increasing function of p_{BC} so that

$$\frac{d^2\tau^A_{AB}}{dp^2_{BC}} \geq 0.$$

Consequently, τ^A_{BC} is maximised when either $p_{BC} = 0$ or $p_{BC} = 1$, and the value of p_{BC} which minimises τ^A_{BC} is obtained by setting the first derivative equal to zero. This completes the proof of Theorem 5.

In this case when $p_{BC} = 0$ then

$$\tau^A_{BC} = \frac{p_{AB}p_{AC}}{1-p_{BA}p_{AC}},$$

and when $p_{BC} = 1$ then

$$\tau^A_{BC} = \frac{p_{AB}p_{AC}}{1-p_{AB}p_{CA}},$$

so that according to Theorem 5 the maximum value of τ^A_{AB} will be the larger of these two values. Interestingly, if $p_{AB} > p_{AC}$ then the larger value occurs when $p_{BC} = 1$, so that A hopes that the contestant that A is strongest against will always beat the contestant that A is weakest against. The minimum value can be calculated to be

$$\tau^A_{BC} = \frac{p_{AB}p_{AC}(p_{BA}p_{AC}+p_{AB}p_{CA})}{p_{BA}p_{AC}+p_{AB}p_{CA}-p_{BA}p_{AC}p_{AB}p_{CA}},$$

and so for fixed p_{AB} and p_{AC} these values provide the range of the possible values of the probability that A wins the tournament after having to sit out the first fight.

2.2 Equally Skilled Contestants

Consider the situation in which $p_{AB} = p_{BC} = p_{CA} = p$. This is a balanced situation in which the three contestants are equivalent to each other. In such a situation, a fair playoff protocol ought to give each contestant a probability of 1/3 of winning the tournament. In particular, the case $p = 1/2$ describes the situation in which the outcomes of each fight are both equally likely.

The probabilities of winning the tournament can be obtained from Theorem 2 to be

$$\tau_{AB}^A = \frac{p(1-p)}{1-p^3} + \frac{p(1-p)^3}{1-(1-p)^3}, \qquad \tau_{AB}^B = \frac{p^3(1-p)}{1-p^3} + \frac{p(1-p)}{1-(1-p)^3},$$

$$\tau_{AB}^C = \frac{p^2(1-p)}{1-p^3} + \frac{p(1-p)^2}{1-(1-p)^3}.$$

These are graphed in Figure 1. When $p = 1/2$ the values are $\tau_{AB}^A = \tau_{AB}^B = 5/14$ with $\tau_{AB}^C = 4/14$, so that the two starting contestants each have winning probabilities 5/4 times larger than the contestant who must sit out the first fight. This is a measure of the unfairness of the playoff protocol.

Notice that as a consequence of Theorem 3, the values of τ_{AB}^A and τ_{AB}^B are always larger than τ_{AB}^C. The largest value of the difference $\max\{\tau_{AB}^A, \tau_{AB}^B\} - \tau_{AB}^C$ occurs when $p = 0.263$ (or equivalently $p = 0.737$) in which case $\tau_{AB}^A = 0.373$, $\tau_{AB}^B = 0.337$ and $\tau_{AB}^C = 0.290$. It is also interesting to notice that $\tau_{AB}^A > \tau_{AB}^B$ when $0 < p < 1/2$ which may at first sight appear counterintuitive.

2.3 Number of Fights Required to Determine the Tournament Winner

Let the random variable N denote the number of fights required to determine the tournament winner. Then the distribution of N depends upon the fight outcome probabilities p_{ij} and the choice of the two contestants in the first fight. The probability mass function of N is given in Theorem 6.

Theorem 6

Suppose that the first fight is between contestants A and B. Then if $c_1 = p_{AB}p_{CA}p_{BC}$ and $c_2 = p_{BA}p_{CB}p_{AC}$, the probability mass function of the number of fights required to determine the tournament winner N is given by

$$P(N = 2 + 3i) = c_1^i p_{AB}p_{AC} + c_2^i p_{BA}p_{BC},$$

$$P(N = 3 + 3i) = c_1^i p_{AB}p_{CA}p_{CB} + c_2^i p_{BA}p_{CB}p_{CA},$$

$$P(N = 4 + 3i) = c_1^{i+1} p_{BA} + c_2^{i+1} p_{AB},$$

for $i = 0, 1, 2, \ldots$.

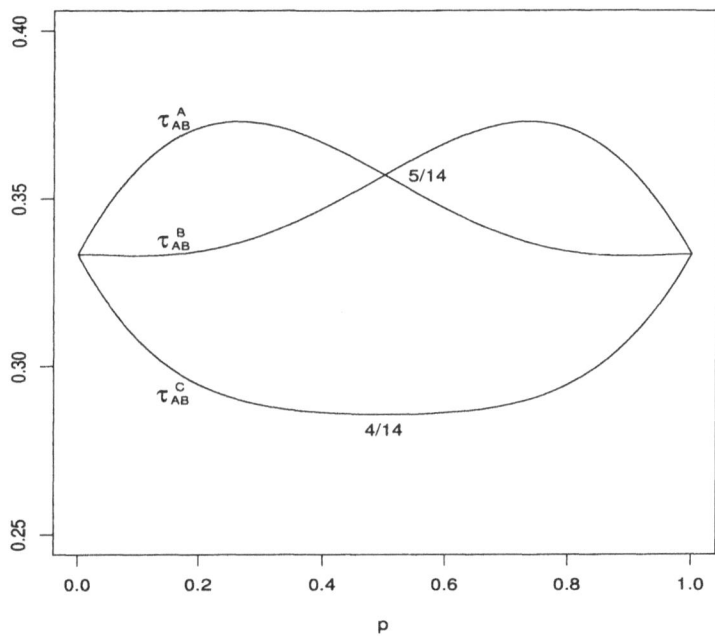

Figure 1: The probabilities of winning the tournament

Proof of Theorem 6

The tournament winner will be decided after exactly two fights if either contestant A or contestant B wins the first two fights, so that

$$P(N = 2) = p_{AB}p_{AC} + p_{BA}p_{BC}.$$

Furthermore, the tournament winner will be decided after exactly three fights if either contestant A or contestant B wins the first fight, and then contestant C wins the second and third fights, so that

$$P(N = 3) = p_{AB}p_{CA}p_{CB} + p_{BA}p_{CB}p_{CA}.$$

Similarly it can be deduced that

$$P(N = 4) = p_{AB}p_{CA}p_{BC}p_{BA} + p_{BA}p_{CB}p_{AC}p_{AB} = c_1 p_{BA} + c_2 p_{AB}.$$

The formulae for larger values of N are derived by considering the cycles of wins and losses which are required to achieve that value of N. If contestant A wins the first fight then the cycle is "A beats B, C beats A, B beats C" which has probability c_1. If contestant B wins the first fight then the cycle is "B beats A, C beats B, A beats C" which has probability c_2. This completes the proof of Theorem 6.

Theorem 7 provides the expected value of the number of fights required to determine the tournament winner.

Theorem 7

Suppose that the first fight is between contestants A and B. Then if $c_1 = p_{AB}p_{CA}p_{BC}$ and $c_2 = p_{BA}p_{CB}p_{AC}$, the expected value of the number of fights required to determine the tournament winner N is

$$E(N) = \frac{p_{AB}(2 + p_{CA} + p_{BA}p_{CA}p_{BC})}{1 - c_1} + \frac{p_{BA}(2 + p_{CB} + p_{AB}p_{CB}p_{AC})}{1 - c_2}.$$

Proof of Theorem 7

Using the probability mass function given in Theorem 6, the expected value of N can be written

$$E(N) = p_{BA} \sum_{j=1}^{\infty}(1 + 3j)c_1^j + p_{AB} \sum_{j=1}^{\infty}(1 + 3j)c_2^j +$$

$$p_{AB}p_{AC} \sum_{j=0}^{\infty}(2 + 3j)c_1^j + p_{BA}p_{BC} \sum_{j=0}^{\infty}(2 + 3j)c_2^j +$$

$$p_{AB}p_{CA}p_{CB} \sum_{j=0}^{\infty}(3 + 3j)c_1^j + p_{BA}p_{CB}p_{CA} \sum_{j=0}^{\infty}(3 + 3j)c_2^j.$$

The given formula for $E(N)$ is then obtained by evaluating the summation terms and simplifying. This completes the proof of Theorem 7.

For the balanced case discussed in section 2.2, the expected value of N is

$$E(N) = \frac{p(2 + p + (1 - p)p^2)}{1 - p^3} + \frac{(1 - p)(2 + (1 - p) + p(1 - p)^2)}{1 - (1 - p)^3}.$$

When $p = 1/2$ this takes the value $E(N) = 3$. Also, when $p = 0.4$ it is $E(N) = 3.167$ and when $p = 0.2$ it is $E(N) = 5.243$.

2.4 A Fair Protocol

A fair playoff protocol can be devised which consists of a sequence of sets of three fights. In each set each contestant fights against each of the other two contestants. As soon as one of the contestants wins both fights within a set, that contestant will be declared to be the tournament winner and the playoff will be finished. The ordering of the three fights within each set is not important from the point of view of the contestants' tournament winning probabilities. However, in order to reduce the number of fights required it is sensible that the winner of the first fight in each set should also participate in the second fight in that set, so that if this contestant wins that second fight also, then the third fight is not required since the tournament winner has already been determined.

The probability that the tournament winner is not decided within a particular set is the probability that each contestant wins one fight and loses one fight, which is

$$c = p_{AB}p_{CA}p_{BC} + p_{BA}p_{CB}p_{AC}.$$

The probability that contestant A, say, wins two fights within a set is $p_{AB}p_{AC}$. Letting t^A represent the probability that contestant A wins the tournament, it can be calculated as the sum of the probabilities of winning in the first set, in the second set, and so on, which is given by

$$t^A = p_{AB}p_{AC} \sum_{i=0}^{\infty} c^i = \frac{p_{AB}p_{AC}}{1-c}.$$

Similarly,

$$t^B = \frac{p_{BA}p_{BC}}{1-c} \quad \text{and} \quad t^C = \frac{p_{CA}p_{CB}}{1-c}.$$

The denominator $1 - c$ is also equal to $p_{AB}p_{AC} + p_{BA}p_{BC} + p_{CA}p_{CB}$.

If $p_{AB} = p_{AC}$ then the value of t^A does not depend upon the value of p_{BC}, and if $p_{AB} > p_{AC}$ then the value of t^A is an increasing function of p_{BC}, and is maximised when $p_{BC} = 1$. Consequently, in general terms, it is to the advantage of contestant A if the opponent against which contestant A is stronger is also as strong as possible against the third contestant. The distribution function of N_f, the

number of fights required to determine the tournament winner, can
be calculated quite easily if for each section some decisions are made
concerning which contestants will participate in the first fight. Notice
that 1, 4, 7,... etc. are not possible values of N_f.

For the balanced situation discussed in section 2.2 with $p_{AB} = p_{BC} = p_{CA} = p$ the tournament winning probabilities are $t^A = t^B = t^C = 1/3$ which is the motivation of a fair protocol. Also, in this case
the distribution function of N_f can be calculated. Specifically,

$$P(N_f = 2) = 2p(1-p) \quad \text{and} \quad P(N_f = 3) = p^2(1-p)+p(1-p)^2 = p(1-p).$$

The probability that the tournament winner is not decided within a
particular set is $c = p^3 + (1-p)^3$, so that more generally

$$P(N_f = 2 + 3i) = (p^3 + (1-p)^3)^i 2p(1-p)$$

and

$$P(N_f = 3 + 3i) = (p^3 + (1-p)^3)^i p(1-p)$$

for $i = 0, 1, 2, \ldots$.

For this balanced case, the expected value of the number of fights
required to determine the tournament winner can be calculated as

$$E(N_f) = \frac{3 - 2p(1-p)}{3p(1-p)}$$

which when $p = 1/2$ takes the value $E(N_f) = 10/3$. Also, when $p = 0.4$
it is $E(N_f) = 7/2$ and when $p = 0.2$ it is $E(N_f) = 5.583$. These values
are larger than the corresponding values for $E(N)$ given at the end
of section 2.3. Consequently, the apparent motivation for the sumo
protocol is that while it does result in unequal tournament winning
probabilities in the balanced case, it requires less fights than the fair
protocol.

2.5 Example

Consider the example in which $p_{AB} = 0.55$ and $p_{AC} = 0.65$ so that
contestant A is stronger than both contestants B and C in individual fights, and is strongest against contestant C. What chance does
contestant A have of winning the tournament?

Suppose that contestant A is chosen to participate in the first fight
against contestant B. In this case the value of τ_{AB}^A decreases monotonically in p_{BC} and takes a maximum value of $\tau_{AB}^A = 0.5849$ when

$p_{BC} = 0$ and a minimum value of $\tau^A_{AB} = 0.4427$ when $p_{BC} = 1$. When $p_{BC} = 0.5$ calculations give $\tau^A_{AB} = 0.4899$ and $E(N) = 2.89$.

Suppose now that contestant A is chosen to participate in the first fight against contestant C. In this case $\tau^A_{AC} = 0.5053$ when $p_{BC} = 0$ and takes a maximum value of $\tau^A_{AC} = 0.5125$ when $p_{BC} = 1$. A minimum value $\tau^A_{AC} = 0.4877$ occurs at $p_{BC} = 0.441$. When $p_{BC} = 0.5$ calculations give $\tau^A_{AC} = 0.4880$ and $E(N) = 2.96$.

Finally, suppose that contestant A is chosen to sit out the first fight. In this case $\tau^A_{BC} = 0.5053$ when $p_{BC} = 0$ which is its maximum value. Also, $\tau^A_{BC} = 0.4427$ when $p_{BC} = 1$, and a minimum value $\tau^A_{BC} = 0.4045$ occurs at $p_{BC} = 0.603$. When $p_{BC} = 0.5$ calculations give $\tau^A_{BC} = 0.4072$ and $E(N) = 3.10$.

In comparison, if the fair protocol described in section 2.4 is used, then the value of t^A decreases monotonically in p_{BC} and takes a maximum value of $t^A = 0.5053$ when $p_{BC} = 0$ and a minimum value of $t^A = 0.4427$ when $p_{BC} = 1$. When $p_{BC} = 0.5$ calculations give $t^A = 0.4719$.

3. Extensions to More than Three Contestants

It is interesting to consider the problem of determining a single winner out of a general number of n contestants when the contestants may fight each other only in pairs. In particular, it is interesting to consider the following extension of the sumo protocol.

Extended Sumo Protocol for n Contestants

Two contestants are chosen to participate in the first fight, and the remaining $n - 2$ contestants form an ordered line waiting to fight. In general, the winner of fight i remains to participate in fight $i + 1$ against the contestant who is at the top of the waiting line. The loser of fight i goes to the end of the waiting line, with everybody else in the waiting line moving up one space. Thus, the loser of fight i will next participate in fight $i + n - 1$, assuming that the tournament has not finished by then. As soon as one contestant wins $n - 1$ consecutive fights (which will necessarily be against each one of the other $n - 1$ contestants) then that contestant is declared the tournament winner and the playoff is finished.

In practice it can be seen that if the contestants are reasonably equal in strength then this protocol has the disadvantage that it may take a large number of fights to determine a tournament winner. Nevertheless, its analysis presents an interesting probabilistic problem. In particular it is interesting to calculate the probabilities of winning the tournament for each of the initial starting positions. In this section this extension of the sumo protocol is studied for the simplest case in which the results of each fight are equally likely with probability $1/2$. In other words, the contestants are all equally skillful.

3.1 General Results

In order to analyse this problem assume that a fight has just ended and the winner of that fight (the incumbent) has had i consecutive wins. Then consider a contestant who is in the waiting line and ask what is the probability that this contestant will win the tournament. This probability will depend upon two things only, which are the value of i and the number of fights j which must take place before the contestant's next turn to fight, and can therefore be denoted by $\pi_j^{(i)}$. The value of i can be between 1 and $n-2$ (if $i = n-1$ then the tournament will have just ended), and the value of j can be between 0 and $n-2$. If $j = 0$ then the contestant will participate in the next fight against the other contestant who already has i consecutive wins. If $j = n-2$ then the contestant has just lost the last fight and is at the end of the waiting line.

Notice that the winner of the last fight who has had i consecutive wins has a probability of winning the tournament which is

$$1 - \sum_{j=0}^{n-2} \pi_j^{(i)},$$

and which can also be expressed as

$$\left(\frac{1}{2}\right)^{n-1-i} + \left(1 - \left(\frac{1}{2}\right)^{n-1-i}\right) \pi_{n-2}^{(1)}$$

since the contestant will either win the tournament by winning the next $n - 1 - i$ fights, or will end up at the back of the line leaving another contestant with one consecutive win. Consequently, the probabilities $\pi_j^{(i)}$ provide a full analysis of this protocol and some important relationships among them are given in Theorem 8.

Theorem 8

The probabilities $\pi_j^{(i)}$ satisfy the following relationships.

$$\pi_0^{(i)} = \left(\frac{1}{2}\right)^{n-1} + \left(\frac{1}{2} - \left(\frac{1}{2}\right)^{n-1}\right)\pi_{n-2}^{(1)} + \frac{1}{2}\pi_{n-2}^{(i+1)} \qquad \text{for } 1 \leq i \leq n-3,$$

$$\pi_0^{(n-2)} = \left(\frac{1}{2}\right)^{n-1} + \left(\frac{1}{2} - \left(\frac{1}{2}\right)^{n-1}\right)\pi_{n-2}^{(1)},$$

$$\pi_j^{(i)} = \frac{1}{2}\pi_{j-1}^{(1)} + \frac{1}{2}\pi_{j-1}^{(i+1)} \qquad \text{for } 1 \leq i \leq n-3,\ 1 \leq j \leq n-2,$$

$$\pi_j^{(n-2)} = \frac{1}{2}\pi_{j-1}^{(1)} \qquad \text{for } 1 \leq j \leq n-2.$$

Proof of Theorem 8

The expression for $\pi_0^{(i)}$ for $1 \leq i \leq n-3$ is calculated as follows. The first term corresponds to the contestant winning the next $n-1$ fights. The second term corresponds to the contestant winning only between one and $n-2$ consecutive fights. The last term corresponds to the contestant losing the next fight. The expression for $\pi_0^{(n-2)}$ is similar except that there is no third term since if the contestant loses the next fight then the playoff is finished and somebody else is the tournament winner.

The expression for $\pi_j^{(i)}$ for $1 \leq i \leq n-3$, $1 \leq j \leq n-2$, can be deduced by considering the two possible outcomes of the fight between the contestant with i consecutive wins and the contestant at the head of the waiting line. Again, when $i = n-2$ the playoff will finish if the contestant with $n-2$ consecutive wins happens to win the next fight also. This completes the proof of Theorem 8.

Define $\boldsymbol{\pi}_j = (\pi_j^{(1)}, \ldots, \pi_j^{(n-2)})'$. Then it follows from Theorem 8 that if one of the vectors $\boldsymbol{\pi}_0, \ldots, \boldsymbol{\pi}_{n-2}$ is known then all of the other vectors can be calculated recursively. Specifically,

$$\boldsymbol{\pi}_{j+1} = A\boldsymbol{\pi}_j \qquad \text{for } 0 \leq j \leq n-3,$$

where the matrix $A = (a_{ij})$ has all entries zero except for $a_{11}, a_{12}, \ldots, a_{1,n-2}$ and $a_{21}, a_{32}, \ldots, a_{n-2,n-3}$ which are all equal to $1/2$. This allows us to write $\boldsymbol{\pi}_{n-2}$ in terms of $\boldsymbol{\pi}_0$ as shown in Theorem 9.

Theorem 9

The values $\pi_{n-2}^{(i)}$ can be written in terms of the values $\pi_0^{(i)}$ in the following manner.

$$\pi_{n-2}^{(1)} = \sum_{r=1}^{n-2} \left(\frac{1}{2}\right)^r \pi_0^{(r)},$$

$$\pi_{n-2}^{(i)} = \left(1 - \left(\frac{1}{2}\right)^{n-1-i}\right) \left(\sum_{r=1}^{i-1} \left(\frac{1}{2}\right)^r \pi_0^{(r)}\right) + \sum_{r=i}^{n-2} \left(\frac{1}{2}\right)^r \pi_0^{(r)}$$

for $2 \le i \le n - 2$.

Proof of Theorem 9

Formally, these equations can be obtained from the expression

$$\boldsymbol{\pi}_{n-2} = A^{n-2}\boldsymbol{\pi}_0.$$

Alternatively, they can be deduced as follows. Consider the contestant whose probability is $\pi_{n-2}^{(1)}$. When this contestant reaches the head of the waiting line, the probability that the contestant will fight a contestant with exactly r consecutive wins is $(1/2)^r$. This is because r fights ago the newcomer must have beaten the incumbent, and that newcomer must have then won the subsequent $r - 1$ fights. This explains the expression for $\pi_{n-2}^{(1)}$ (there is also a probability of $(1/2)^{n-2}$ that the playoff finishes just as the contestant reaches the head of the waiting line, but this does not enter into the expression).

Now for $2 \le i \le n - 2$ consider the contestant whose probability is $\pi_{n-2}^{(i)}$. The terms in the expression for $\pi_{n-2}^{(i)}$ can be argued in a similar manner as for $\pi_{n-2}^{(1)}$ except that for $1 \le r \le i - 1$ it must be ensured that the playoff has not finished due to the current incumbent with i consecutive wins also winning the next $n - 1 - i$ fights. This is accounted for in the factor

$$\left(1 - \left(\frac{1}{2}\right)^{n-1-i}\right).$$

This completes the proof of Theorem 9.

Notice that while Theorem 9 provides equations for the values $\pi_{n-2}^{(i)}$ in terms of the values $\pi_0^{(i)}$, the first two lines of Theorem 8 provide equations for the values $\pi_0^{(i)}$ in terms of the values $\pi_{n-2}^{(i)}$. Together, these equations can be solved to find the values of $\pi_0^{(i)}$ and $\pi_{n-2}^{(i)}$. Theorem 10 provides the values of $\pi_0^{(i)}$.

Theorem 10

The values of $\pi_0^{(1)}, \ldots, \pi_0^{(n-2)}$ are given by

$$\pi_0^{(i)} = \frac{2^{n-2}\left(2^{n-1} + 1 - 2^i \left(\frac{2^{n-1}}{1+2^{n-1}}\right)^{i-1}\right)}{(4^{n-1} - 1)\left(2^{n-2} + 1 - 2^{n-2}\left(\frac{2^{n-1}}{1+2^{n-1}}\right)^{n-2}\right)}$$

for $1 \leq i \leq n - 2$.

Proof of Theorem 10

If the expressions for $\pi_{n-2}^{(i)}$ in terms of the values $\pi_0^{(i)}$ given in Theorem 9 are substituted into the expressions for $\pi_0^{(i)}$ in terms of the values $\pi_{n-2}^{(i)}$ given in the first two lines of Theorem 8, then the equations

$$\pi_0^{(i)} = \left(\frac{1}{2}\right)^{n-1} + \left(1 - \left(\frac{1}{2}\right)^{n-1} - \left(\frac{1}{2}\right)^{n-1-i}\right)\left(\sum_{r=1}^{i}\left(\frac{1}{2}\right)^r \pi_0^{(r)}\right)$$

$$+ \left(1 - \left(\frac{1}{2}\right)^{n-1}\right)\sum_{r=i+1}^{n-2}\left(\frac{1}{2}\right)^r \pi_0^{(r)}$$

for $1 \leq i \leq n - 2$ are obtained. It then only remains to check that these equations are satisfied by the values for $\pi_0^{(1)}, \ldots, \pi_0^{(n-2)}$ given in this theorem, which will complete the proof of Theorem 10.

As explained before, these values for π_0 allow the calculation of the remaining probabilities $\pi_1, \ldots, \pi_{n-2}$ using the relationship $\pi_{j+1} = A\pi_j$. Thus they provide a complete description of the probabilistic properties of this protocol.

Specifically, the probabilities π_0 can be used to calculate the tournament winning probabilities for each of the starting positions. In this respect define $\tau_1, \ldots, \tau_n$ to be the tournament winning probabilities for starting positions $1, \ldots, n$, where $\tau_1 = \tau_2$ are the tournament winning probabilities for the two contestants who participate in the first fight, and for example τ_3 is the tournament winning probability for the contestant who starts at the head of the waiting line while τ_n is the tournament winning probability for the contestant who starts at the end of the waiting line. Of course $\tau_1 + \cdots + \tau_n = 1$. Also, $\tau_i = \pi_{i-3}^{(1)}$. These tournament winning probabilities are given in Theorem 11.

Theorem 11

The tournament winning probabilities $\tau_1, \ldots, \tau_n$ for each of the starting positions are given by

$$\tau_i = \frac{1 - \left(\frac{2^{n-1}}{1+2^{n-1}}\right)}{2 - \left(\frac{2^{n-1}}{1+2^{n-1}}\right) - \left(\frac{2^{n-1}}{1+2^{n-1}}\right)^{n-1}} \left(\frac{2^{n-1}}{1+2^{n-1}}\right)^{i-2} \qquad \text{for } 2 \leq i \leq n$$

with $\tau_1 = \tau_2$.

Proof of Theorem 11

Notice that $\tau_3 = \pi_0^{(1)}$. Also, when the contestant in starting position 4 reaches the head of the waiting line the incumbent will have one consecutive win with probability $1/2$ or will have two consecutive wins with probability $1/2$. Therefore

$$\tau_4 = \frac{1}{2}\pi_0^{(1)} + \frac{1}{2}\pi_0^{(2)}.$$

More generally, consider τ_i for $4 \leq i \leq n$. The contestant in starting position i will certainly reach the head of the waiting line before the playoff is finished. At that point the incumbent will have j consecutive wins with a probability $(1/2)^j$ for $1 \leq j \leq i-3$, and also will have $i-2$ consecutive wins with a probability $(1/2)^{i-3}$. Therefore

$$\tau_i = \sum_{j=1}^{i-3} \left(\frac{1}{2}\right)^j \pi_0^{(j)} + \left(\frac{1}{2}\right)^{i-3} \pi_0^{(i-2)}$$

for $4 \leq i \leq n$. Consequently, the values of $\tau_3, \ldots, \tau_n$ can be calculated from the values of π_0 given in Theorem 10. The values of $\tau_1 = \tau_2$ can then be calculated using the fact that the τ_i all sum to one. This completes the proof of Theorem 11.

Notice that when $n = 3$ the tournament winning probabilities are $\tau_1 = \tau_2 = 5/14$ and $\tau_3 = 4/14$ as shown in section 2.2. The cases $n = 4$ and $n = 5$ are considered in sections 3.3 and 3.4 respectively.

In general notice that the probabilities $\tau_2, \ldots, \tau_n$ form a decreasing geometric series defined by

$$\tau_i = \left(1 + \left(\frac{1}{2}\right)^{n-1}\right) \tau_{i+1}.$$

Thus the two contestants who participate in the first fight have the greatest chance of winning the tournament, and the chances decrease the further down the initial waiting line a contestant starts. The ratio of the largest tournament winning probability to the smallest tournament winning probability is

$$\frac{\tau_2}{\tau_n} = \left(1 + \left(\frac{1}{2}\right)^{n-1}\right)^{n-2}$$

which is maximised at $n = 4$ when it takes the value $81/64 = 1.266$. Also, as $n \to \infty$ all of the τ_i tend to $1/n$.

3.2 Number of Fights Required to Determine the Tournament Winner

Let the random variable N be the number of fights required to determine the tournament winner. Theorem 12 provides the expected value of N.

Theorem 12

For $1 \leq i \leq n-2$ define m_i to be the expected number of additional fights needed to determine the tournament winner when the current incumbent contestant has i consecutive wins. Then

$$m_i = 2^{n-1} - 2^i.$$

Also, $E(N) = 1 + m_1 = 2^{n-1} - 1$.

Proof of Theorem 12

The consideration of the two possible outcomes of the next fight leads to the following recursive relationships among the m_i,

$$m_i = 1 + \frac{1}{2}m_1 + \frac{1}{2}m_{i+1} \quad \text{for } 1 \leq i \leq n - 3$$

$$m_{n-2} = 1 + \frac{1}{2}m_1,$$

which are satisfied by $m_i = 2^{n-1} - 2^i$. This completes the proof of Theorem 12.

Notice that when $n = 3$ the expected value of the number of fights required to determine the tournament winner is $E(N) = 2^{3-1} - 1 = 3$ as discussed in section 2.3.

The actual distribution function of the random variable N is rather diffcult to write down explicitly for general n. However, the values of the distribution function can be obtained recursively from the relationship given in Theorem 13.

Theorem 13

Let $f_i = P(N = i)$ for $i \geq n - 1$. Then

$$f_{n-1} = \left(\frac{1}{2}\right)^{n-2}$$

$$f_i = \left(\frac{1}{2}\right)^{n-1} \qquad \text{for } n \leq i \leq 2n - 3$$

$$f_i = \left(1 - \sum_{j=n-1}^{i-n+1} f_j\right) \left(\frac{1}{2}\right)^{n-1} \qquad \text{for } i \geq 2n - 2.$$

Proof of Theorem 13

Notice that f_{n-1} is the probability that either of the two contestants who participate in the first fight actually win the first $n-1$ fights, which is $(1/2)^{n-2}$. Furthermore, for $n \leq i \leq 2n - 3$, f_i is the probability that the contestant who was in position $i - n + 1$ in the initial waiting line wins fights $i - n + 2$ to i, which is $(1/2)^{n-1}$. For $i \geq 2n - 2$, $N = i$ if the newcomer beats the incumbent in fight $i - n + 2$ and then wins the next $n - 2$ fights as well, which has a probability $(1/2)^{n-1}$. However, to calculate f_i, this probability needs to be multiplied by the probability that the playoff has not finished before fight $i - n + 2$, which is $1 - \sum_{j=n-1}^{i-n+1} f_j$. This completes the proof of Theorem 13.

From Theorem 13 the values $P(N = n - 1), \ldots, P(N = 2n - 3)$ are immediately known, and the subsequent values of the distribution function of N can be obtained successively from the recursive formula. When $n = 3$ the solutions to the equations given in Theorem 13 are $P(N = i) = (1/2)^{n-1}$ which is a special case of Theorem 6. When $n = 4$ the solutions to the equations given in Theorem 13 are shown in section 3.3.

3.3 Four Contestants

When $n = 4$ the following results can be obtained. At each stage of the playoff, the winning probabilities are

$$\pi_0 = \left(\frac{36}{149}, \frac{28}{149}\right), \pi_1 = \left(\frac{32}{149}, \frac{18}{149}\right), \pi_2 = \left(\frac{25}{149}, \frac{16}{149}\right).$$

Also, a contestant with one consecutive win has a probability of $56/149 = 37.58\%$ of winning the tournament, and a contestant with two consecutive wins has a probability of $87/149 = 58.39\%$ of winning the tournament. The probabilities of winning the tournament for the various starting positions are

$$\tau_1 = \tau_2 = \frac{81}{298} = 27.18\%, \tau_3 = \frac{72}{298} = 24.16\%, \tau_4 = \frac{64}{298} = 21.48\%,$$

and as mentioned at the end of section 3.1, the ratio $\tau_4/\tau_1 = 1.266$ is the largest value of τ_n/τ_1 over all n. Finally, the distribution function of N is

$$P(N = i) = \frac{1}{2\sqrt{5}} \left(\left(\frac{1}{4} + \frac{\sqrt{5}}{4} \right)^{i-2} - \left(\frac{1}{4} - \frac{\sqrt{5}}{4} \right)^{i-2} \right)$$

with $E(N) = 7$.

3.4 Five Contestants

When $n = 5$ the following results can be obtained. At each stage of the playoff, the winning probabilities are

$$\pi_0 = \left(\frac{2312}{11449}, \frac{2040}{11449}, \frac{1528}{11449} \right), \pi_1 = \left(\frac{2176}{11449}, \frac{1920}{11449}, \frac{1156}{11449} \right),$$

$$\pi_2 = \left(\frac{2048}{11449}, \frac{1666}{11449}, \frac{1088}{11449} \right), \pi_3 = \left(\frac{1857}{11449}, \frac{1568}{11449}, \frac{1024}{11449} \right).$$

Also, a contestant with one consecutive win has a probability of $3056/11449 = 26.69\%$ of winning the tournament, a contestant with two consecutive wins has a probability of $4255/11449 = 37.16\%$ of winning the tournament, and a contestant with three consecutive wins has a probability of $6653/11449 = 58.11\%$ of winning the tournament. The probabilities of winning the tournament for the various starting positions are

$$\tau_1 = \tau_2 = \frac{4913}{22898} = 21.46\%, \tau_3 = \frac{4624}{22898} = 20.19\%,$$

$$\tau_4 = \frac{4352}{22898} = 19.01\%, \tau_5 = \frac{4096}{22898} = 17.89\%.$$

Finally, $E(N) = 15$.

References

Bechhofer, R. E. (1970). On ranking the players in a 3-player tournament. *Nonparametric Techniques in Statistical Inference*, edited by Puri, M. L., 545-559, Cambridge University Press.

Brace, A. and Brett, J. (1975). An alternative to the round robin tournament. *Combinatorial Mathematics, III*, 62-78, Lecture Notes in Mathematics, Volume 452, Berlin: Springer.

Sobel, M. and Weiss, G. (1970). Inverse sampling and other selection procedures for tournaments with two or three players. *Nonparametric Techniques in Statistical Inference*, edited by Puri, M. L., 515-543, Cambridge University Press.

Quantification of Ordinal Variables:

A Critical Inquiry into

Polychoric and Canonical Correlation

Shizuhiko Nishisato[1] and David Hemsworth[2]

[1] The Ontario Institute for Studies in Education of the University of Toronto
252 Bloor Street, Toronto, Ontario, CANADA M5S 1V6
[2] Wilfrid Laurier University, Waterloo, Ontario, CANADA N2L 3C5

Summary: "Scaling" or "quantification" was re-examined with respect to its main objectives and requirements. Among other things, the attention was directed to the condition that the variance-covariance matrix of scaled quantities must be positive definite or semi-definite in order for the variables to be mapped in Euclidean space. Dual scaling was chosen to guide us through the search for identifying problems, understanding the basic aspects of those problems and practical remedies for them. In particular, the pair-wise quantification approach and the global quantification approach to multivariate analysis were used to identify some tricky theoretical problems, associated with the failure of identifying coordinates of variables in Euclidean hyperspace. One of the problems, arising from the pair-wise quantification, was the lack of a geometric definition of correlation between two sets of categorical variables. This absence of a geometric definition was attributed to the lack of a single data matrix, often leading to negative eigenvalues of the correlation matrix. Then the attention was shifted to the calculation of polychoric correlation and canonical correlation for categorized ordinal variables, the practice often seen in the study of structural equation modeling (SEM). Of particular interest were the problems associated with the pair-wise determination of thresholds (polychoric correlation), the univariate determination of thresholds (polychoric correlation) and the pair-wise determination of category weights (canonical correlation). The study identified two possible causes for the failure of mapping variables in Euclidean space : the pair-wise determination of thresholds or categories and the lack of underlying multivariate normality of the distribution. The degree of this failure was noted as an increasing function of the number of variables in the data set. It was highlighted then that dual scaling could mitigate the problems due to these causes that the current SEM practice of using polychoric correlation and canonical correlation would constantly encounter. Numerical examples were provided to show what is at stake when scaling is not properly carried out. It was stressed that when one cannot reasonably make the assumption of the latent multivariate normal distribution dual scaling offers an excellent alternative to canonical correlation and polychoric correlation as used in SEM because dual scaling transforms the data towards the categorized normal distribution.

1. Introduction

Scaling is one of the major branches of psychometrics. It can be characterized as a family of transformations by which the level of input measurement is upgraded. In Stevens' terminology (Stevens, 1951), "measurement" is defined as assignment of numbers to objects according to certain rules, and measurement can be classified into four categories: the lowest level is

1. *Nominal Measurement* where the only rule is the one-to-one correspondence (e.g., back numbers such as 1, 3 and 16 used for baseball players as labels). The next level is

2. *Ordinal Measurement* where the rules are monotone transformation as well as the one-to-one correspondence (e.g., ranks of movie stars' popularity). The next level is

3. *Interval Measurement* where the rules are the equality of the unit and the two rules for ordinal measurement (e.g., temperature measurement in Celsius). The highest level is

4. *Ratio Measurement* where the rules include those three for interval measurement and the rational origin (e.g., measurement of distance in meters, where '0' is defined as 'nothingness' of the distance).

Thus, scaling of nominal measurement is expected to produce at least ordinal measurement, or hopefully interval or ratio measurement. This process of upgrading of the levels of measurement is the main object of scaling. The outcome of scaling is therefore that scaled measurement should be amenable to more algebraic operations than the original measurement. Thus, the main realm of scaling is largely restricted to categorical data analysis. Some of the well-known scaling procedures are:

1. Thurstone-Bradley-Terry-Luce models for paired comparisons (Thurstone, 1927; Bradley and Terry, 1952; Luce, 1959) to produce interval measurement from ordinal measurement,

2. Coombs' unfolding technique (Coombs, 1964) to transform ordinal measurement to interval measurement,

3. Hayashi's theory of quantification (Hayashi, 1950) to generate interval measurement from nominal or ordinal measurement,

4. Guttman's scalogram analysis (Guttman, 1950) to produce ordinal measurement from nominal measurement, and

5. Shepard-Kruskal non-metric multidimensional scaling (Shepard, 1962a, 1962b; Kruskal, 1964a, 1964b) to transform ordinal measurement to interval measurement.

Most of these scaling methods are not new and are taken for granted as if no further work on them is needed. In the advent of the computer age, it is timely to reflect on some of the fundamental aspects of scaling. For one may be tempted to subject scaled outcomes from a computer program to further statistical analysis without any deliberate attempts to assess the quality of their derived measurement. An immediate consequence is inferred from the adage "garbage in garbage out".

Recently, it came to our attention that a widely used computer program provides at best a questionable measurement for data processing. This problem was a surprising discovery, which has obviously been treated by many researchers simply as problematic, rather than detrimental or illogical. The current paper was motivated by that observation, and therefore aims to convey messages of warning that somehow the serious intention behind scaling has been compromised, thus resulting as if in building the tower of Babel in the scientific community of cumulative disciplines.

In this paper, we will first look at some basic aspects of scaling, then dual scaling and its promising features for data analysis. With these preliminaries, the main notions of pair-wise quantification and global quantification will be presented. They will then be portrayed as two distinct approaches to scaling, one leading to serious problems and the other as a justifiable approach in line with the basic aims of scaling. In this context, we will look at the current practices of calculating canonical correlation and polychoric correlation for categorized ordinal variables in SEM as examples of misapplied scaling problems, and possible remedies will be presented to conclude the paper with some bold conjectures.

2. Fundamental Requirements of Scaling

As mentioned above, scaling is mainly used to upgrade the level of measurement. One of the fundamental requirements for scaling is that upgraded measurement should be amenable to more arithmetic operations. How can we ascertain that this requirement is met? This alone is too large a topic for the current paper, but it should nevertheless be kept in mind as an important problem. As another requirement, we would like to be certain that derived numerals can be understandable within our common-sense framework. Considering that the task of scaling mostly involves multivariate data, the derived measurement should satisfy what the Young-Householder theorem (1938) ascertains.

2.1 Young-Householder Theorem

1. If an $n \times n$ scalar-product matrix $\mathbf{A}$ of scaled measurements is positive semi-definite, the distance between any two points may be considered as distance between points lying in a real Euclidean space.

2. The rank of any positive semi-definite matrix $\mathbf{A}$ is equal to the dimensionality of the space required to represent n points.

3. Any positive semi-definite matrix $\mathbf{A}$ may be factored to obtain a matrix $\mathbf{B}$, where $\mathbf{A} = \mathbf{BB}'$. If the rank of matrix $\mathbf{A}$ is K, then $\mathbf{B}$ is a matrix of coordinates of n points on K dimensions in Euclidean space.

This theorem became popular when Torgerson (1952) published his pioneering paper on metric multidimensional scaling. His main task was to obtain a matrix of coordinates for n variables, given an n-by-n matrix of inter-variable distances. The first problem he stumbled on was that psychological data of inter-variable dissimilarities would not yield distance measurements, that is, ratio measurements but only interval measurements. The estimation of the origin of measurement therefore became the key issue for him to derive distance measurements from interval measurements. In other words, he hoped to estimate the origin of measurement so that interval measurement could be converted to ratio measurement in the process of multidimensional scaling.

Some researchers (e.g., Messick and Abelson, 1956) decided to determine this unknown origin by requiring that the variance-covariance matrix of dissimilarity measurements (plus an unknown constant) be positive definite or positive semi-definite. They considered this requirement as the criterion in determining the unknown constant, and tried to derive distance measurements. Historically, this is referred to as *the additive constant problem* in multidimensional scaling. In practice, it is well-known in matrix algebra that given a matrix of real numbers, say $\mathbf{X}$, the cross product matrix $\mathbf{X'X}$ or $\mathbf{XX'}$ is always positive definite or semi-definite. This suggests that once we have a data matrix, or a matrix of scaled data, it is guaranteed that we will have a positive definite or semi-definite correlation matrix. This is a simple matter to realize, but for some reasons or others this fundamental and simple fact seems to have been forgotten by some researchers.

In retrospect, it was remarkable that such researchers as Messick and Ableson had chosen the positive-definiteness as the criterion for determining the positive constant for multidimensional scaling. It was a laudable and probably most sensible criterion one could ever consider. It assures us that we are indeed dealing with the decomposition of data in Euclidean space. Surprisingly, however, many researchers today do not appear to pay much attention to this fundamental requirement for measurement mainly because a computer-intensive method of nonmetric multidimensional scaling, which does not require ratio measurement, has gained the popularity.

2.2 Metric Axioms

There is another set of rules that scaling attempts to satisfy, metric axioms. A scaled measurement of distance d_{jk} between point j and point k is said to be metric if:

1. $d_{jj} = 0$, that is, the distance from point j to itself is zero.

2. $d_{jk} \geq 0$, that is, the distance is non-negative.

3. $d_{jk} = d_{kj}$, that is, the distance is symmetric.

4. $d_{ik} \leq d_{ij} + d_{jk}$: triangular inequality: One segment of a triangle cannot be greater (longer) than the sum of the remaining segments.

5. That $d_{jk} = 0$ means $j = k$.

There are some measures that do not satisfy all the metric axioms. For instance, d_{jk} is said to be *pseudometric* if it satisfies (1) to (4), *semi-metric* if (1), (2), (3) and (5) hold for it, and *semi-pseudometric* if only (1), (2) and (3) are true.

From the scaling point of view, we would like to have metric measurement that satisfies all five conditions. Distance defined in the Euclidean space satisfies all the axioms.

What will happen if some of the metric axioms are not satisfied? Let us consider an example in which the triangular inequality is not met and a serious consequence of this failure for data analysis. Consider the unbiased estimate of the variance of variable X, which is given by

$$s^2 = \frac{\mathbf{x'x} - \mathbf{x'Px}}{N-1}, \qquad \mathbf{P} = \mathbf{1}(\mathbf{1'1})^{-1}\mathbf{1'} \qquad (2.1)$$

where $\mathbf{x}$ is an $N \times 1$ vector of observations, N the number of subjects and $\mathbf{1}$ is the $N \times 1$ vector of 1's and $\mathbf{P}$ is the projector for the mean subspace. In Euclidean space, the norm of $\mathbf{Px}$ cannot be larger than the norm of $\mathbf{x}$.

$$\| \mathbf{x'x} \| \geq \| \mathbf{x'Px} \| \qquad (2.2)$$

This can easily be seen by noting that $\mathbf{x'x} = \mathbf{x'Px} + \mathbf{x'}(\mathbf{I} - \mathbf{P})\mathbf{x}$ and that all the three terms are quadratic forms and thus positive. If the triangular inequality is not satisfied, however, the above relation is not likely to be satisfied either, and it is possible that the projected vector **Px** may be longer than the original vector **x**. In this case, it follows that the unbiased estimate of the variance becomes negative, which is obvious from the numerator of the equation for s^2. What happens there is that we are dealing with a curved space in which the distance of one segment in a triangle becomes longer than the sum of the remaining two segments. Try to consider three points in a two-dimensional graph and connect two points, not by a straight line but a curved line so that the tri-angular inequality no longer holds.

In examining the metric axioms, it is useful to know a family of metrics which is generally referred to as the Minkowski power metric (Minkowski, 1896). This metric became very popular in the literature on multidimensional scaling in the 1960s.

2.3 The Minkowski Power Metric

The Minkowski power metric $d_{jk}^{(p)}$ is given by

$$d_{jk}^{(p)} = (\sum_{i=1}^{K} |d_{ji} - d_{ki}|^p)^{\frac{1}{p}}. \tag{2.3}$$

When $p \geq 1$, it satisfies all the metric axioms. It is not necessay that p is an integer. In other words, p of 2.68 in the above formula will produce a metric that satisfies the metric axioms. There are some special designations of metrics for particular values of p. When $p = 1$ and the variables are binary, it is called the *Hamming distance* (Hamming, 1950). When $p = 1$ and the variables are continuous, it is called the *city-block metric* or the *Manhattan metric* (Torgerson, 1958). When $p = 2$, it yields the *Euclidean distance*. The Euclidean distance is the most common to all of us, and it is definitely a preferred metric over the others because the distance between any two points in Euclidean space remains invariant over the orthogonal rotation of axes. This is a very important and useful property for multivariate analysis. In most cases, scaling aims to produce the Euclidean distance. Keep in mind that when $p < 1$, the metric axioms do not hold, and a strange thing will happen such as the variance, based on the squared quantity, becomes negative.

3. Case of Two Categorized Variables

In this paper, we consider two cases, one that of two categorized variables and the other that of more than two categorized variables.

3.1 Dual Scaling

Dual scaling, coined by Nishisato (1980), is a quantification method, which has many aliases such as the method of reciprocal averages (Richardson and Kuder, 1933; Horst, 1935), simultaneous linear regressions (Hirschfeld, 1935; Lingoes, 1964), Hayashi's theory of quantification (Hayashi, 1950, 1952), principal component analysis of categorical data (Torgerson, 1958), optimal scaling (Bock, 1960), analyse des correspondances (Escofier-Cordier, 1969; Benzécri and Cazes, 1973), correspondence analysis (Hill, 1974), and homogeneity analysis (Gifi, 1990). It is based on singular-value decomposition (e.g., Beltrami, 1873; Jordan, 1874; Schmidt, 1907) of not continuous variables but of categorical variables. In dual scaling, these categorical variables are further divided into incidence data and dominance data (Nishisato, 1993) because of the two respective objectives that govern dual scaling (Nishisato, 1996).

Possible reasons for so many aliases are first because there are many ways to formulate this quantification problem and second because the idea of quantification of categorical data was conceived by many researchers in different disciplines and different countries without mutual communication (Nishisato, 1980). To name a few of the approaches to this quantification, Nishisato (1980) discusses the one-way analysis of variance approach, the bilinear correlation approach, the simultaneous linear regression approach, the canonical correlation approach and the reciprocal averaging approach.

All the approaches are based on the same mathematical structure, that is, singular value decomposition of categorical data. Given a two-way table of data with typical element f_{ij}, which may be a frequency or binary (1, 0) element, singular-value decomposition of categorical data can be described as a bilinear decomposition formula,

$$f_{ij} = \frac{f_{i.}f_{.j}}{f_{..}} \{1 + \rho_1 y_{i1} x_{j1} + \rho_2 y_{i2} x_{j2} + \cdots + \rho_K y_{iK} x_{jK}\}, \qquad (3.1)$$

where ρ_k is the k-th largest singular value, y_{ik} is the i-th element of singular vector $\mathbf{y_k}$ for the rows, x_{jk} is the j-th element of singular vector $\mathbf{x_k}$ for the columns of the table, and $K+1$ is the rank of the data matrix. These singular vectors can be viewed as the weight vectors for the rows and the columns.

An important point of this decomposition formula is that it is expressed as a function of singular values. In terms of dual scaling, those vectors for the rows and the columns are referred to as the *optimal* vector of weights for the rows and the *optimal* vector for the columns, respectively. When categorical variables are weighted by these weights, the correlation between the data weighted by $\mathbf{y_k}$ and the data weighted by $\mathbf{x_k}$ are conditionally maximized, and the maximized correlation is nothing but ρ_k. This is what dual scaling accomplishes, and its outputs are the set $(\rho_k, \mathbf{y_k}, \mathbf{x_k})$.

The bilinear expression in terms of singular values can be shown to be mathematically equivalent to the following expressions (Nishisato, 1980):

$$(a) : \mathbf{F} = \frac{1}{f_{..}} \mathbf{D_r Y' \Lambda X D_c}, \qquad (3.2)$$

where $\mathbf{F} = (f_{ij})$, $\mathbf{Y} = (\mathbf{1}, \mathbf{y_1}, \mathbf{y_2}, \ldots, \mathbf{y_K})$, $\mathbf{X} = (\mathbf{1}, \mathbf{x_1}, \mathbf{x_2}, \ldots, \mathbf{x_K})$, $\mathbf{D_r} = \mathrm{diag}(f_{i.})$, $\mathbf{D_c} = \mathrm{diag}(f_{.j})$, and $\mathbf{\Lambda} = \mathrm{diag}(1, \rho_1, \rho_2, \ldots, \rho_K)$. The above is a direct translation of the bilinear expression in matrix terms.

$$(b1) : \rho_k \mathbf{y} = \mathbf{D_r^{-1} F x}, \qquad (3.3)$$

$$(b2) : \rho_k \mathbf{x} = \mathbf{D_c^{-1} F' y}. \qquad (3.4)$$

This set of formulas is called *dual relations* (Nishisato, 1980), or *transition formulas* (Benzécri and Cazes, 1973). It suggests a computational scheme, called the *method of reciprocal averages* (Richardson and Kuder, 1933; Horst, 1935). Suppose we have two multiple-choice questions:

Q1 - "How often do you have nightmares?" Never (1), Sometimes (2), Often (3), Every night (4);

Q2 - "How strongly do you object to the use of sleeping pills?" Very strongly (1), moderately strongly (2), mildly (3), not at all (4).

We can ask how appropriate those weights in the parentheses are, that is, 1, 2, 3 and 4 for four ordered categories (these weights are often referred to as Likert scores (Likert, 1932)). There is a simple way to check the appropriateness of those weights (Nishisato, 1980). Once you obtain the data in the form of a contingency table of categories of Q1 and categories of Q2. Using the weights for the categories of Q2, say u_i, calculate the mean values of the categories of Q1, say $M_{x(j)}$, that is,

$$M_{x(j)} = \frac{\sum_i f_{ij} u_i}{\sum_i f_{.j}}. \tag{3.5}$$

Similarly, using the weights for the categories of Q1, say v_j, calculate the mean values of the categories of Q2, say $M_{y(i)}$, that is,

$$M_{y(i)} = \frac{\sum_j f_{ij} v_j}{\sum_j f_{i.}}. \tag{3.6}$$

Plot those means against the initial category weights (i.e., $M_{y(i)}$ against u_i, and $M_{x(j)}$ against v_j). The closeness of each of the two lines to linearity indicates appropriateness of the original category weights. If we now replace the original category weights with the mean values (now call them category weights) and re-calculate the new mean values, and plot the means against the category weights, you will see that the two lines are straighter than the previous lines. This process is mathematically convergent, and eventually the two lines merge into a single line with the slope equal to the singular value. This is the essence of the *method of reciprocal averages* and *simultaneous linear regressions*. The final category weights are optimal weights.

$$(c1): \rho_k^2 = \frac{\mathbf{y}'\mathbf{F}\mathbf{D}_c^{-1}\mathbf{F}'\mathbf{y}}{\mathbf{y}'\mathbf{D}_r\mathbf{y}}, \tag{3.7}$$

$$(c2) : \rho_k^2 = \frac{\mathbf{x'F'D_r^{-1}Fx}}{\mathbf{x'D_cx}}. \qquad (3.8)$$

These are correlation ratios calculated from the data weighted by the row weights and weighted by the column weights, respectively. The task of dual scaling is to determine $\mathbf{y}$ and $\mathbf{x}$ such that maximize the correlation ratio. In practice, we can use Lagrange's method of an unknown multiplier, and solve the following equations,

$$\frac{\partial}{\partial \mathbf{y}}(\mathbf{y'FD_c^{-1}F'y} - \lambda(\mathbf{y'D_ry} - c)) = \mathbf{0}, \qquad (3.9)$$

$$\frac{\partial}{\partial \mathbf{x}}(\mathbf{x'F'D_r^{-1}Fx} - \lambda(\mathbf{x'D_cx} - c)) = \mathbf{0}, \qquad (3.10)$$

where c is a scaling constant, and λ is the eigenvalue, which can be shown to be equal to the correlation ratio or the squared singular value, that is, ρ^2. These equations lead to the generalized eigen-equations,

$$(\mathbf{FD_c^{-1}F'} - \lambda \mathbf{D_r})\mathbf{y} = \mathbf{0}, \qquad (3.11)$$

$$(\mathbf{F'D_r^{-1}F} - \lambda \mathbf{D_c})\mathbf{x} = \mathbf{0}. \qquad (3.12)$$

The two expressions provide the identical value of correlation ratio, and lead to the optimal weight vectors $\mathbf{x}$ and $\mathbf{y}$.
Because the three formulas (a), (b) and (c) are all expressed in association with singular values, the corresponding vectors (matrices) of weights are all optimal. For details, see, for example, Nishisato (1980, 1994), Greenacre (1984) and Gifi (1990).
The squared singular values, ρ_k^2, are equivalent to the eigenvalues, correlation ratios and squared product-moment correlations between

data weighted by row weights and those by column weights. As such, it is referred to as the information of the components (Nishisato, 1996). For the $m \times n$ contingency table, generated from two categorical variables, the total information is given by

$$\sum_{k=1}^{K} \rho_k^2 = \frac{\chi^2}{f_{..}}.$$
(3.13)

3.2 The Kendall-Stuart Canonical Correlation

The criterion of maximizing correlation in terms of weight vectors has a profound implication for the topic to be discussed here. The link between the topic and the optimization process can be found in the famous paragraph from Kendall and Stuart (1961). They considered the correlation between two categorized variables, say $N \times p$ incidence matrix $\mathbf{F_1}$ and $N \times q$ incidence matrix $\mathbf{F_2}$, the task being to determine a $p \times 1$ weight vectors $\mathbf{w_1}$ for $\mathbf{F_1}$ and a $q \times 1$ weight vector $\mathbf{w_2}$ for $\mathbf{F_2}$ such that the correlation between $\mathbf{F_1 w_1}$ and $\mathbf{F_2 w_2}$ be a maximum. Let us indicate by $\mathbf{D_1}$ the diagonal matrix of column totals of $\mathbf{F_1}$ and by $\mathbf{D_2}$ that of $\mathbf{F_2}$. Suppose we scale the data in such a way that the average of each of the weighted data is zero. Then the correlation we are interested in can be expressed as

$$r = \frac{\mathbf{w_1 F_1' F_2 w_2}}{\sqrt{\mathbf{w_1' D_1 w_1 w_2' D_2 w_2}}}.$$
(3.14)

The Lagrangian function for this optimization can be stated as

$$Q(\mathbf{w_1}, \mathbf{w_2}, \lambda_1, \lambda_2) = \mathbf{w_1' F_1' F_2 w_2} \\ - \frac{1}{2}\lambda_1(\mathbf{w_1' D_1 w_1} - c) - \frac{1}{2}\lambda_2(\mathbf{w_2' D_2 w_2} - c).$$
(3.15)

Partially differentiating Q with respect to two Lagrangian multipliers λ_1 and λ_2 and setting the partial derivatives equal to $\mathbf{0}$, we can find the optimal weight vectors $\mathbf{w_1}$ and $\mathbf{w_2}$ and the canonical correlation r (see Nishisato, 1980). And this is the same as dual scaling of two categorical variables.

This is referred here as the Kendall-Stuart canonical correlation. They state that

> "If we seek separate scoring systems for the two categorized variables such as to maximize their correlation, we are basically trying to produce a bivariate normal distribution by operation upon the margins of the table" (Kendall and Stuart, 1961, Page 569).

The above statement is based on the fact that product-moment correlation is a maximum when the two variables follow a bivariate normal distribution. Because dual scaling is a method to decompose data in terms of singular values, the Kendall-Stuart statement offers a concise description of what dual scaling (Nishisato, 1980, 1994) does for two categorized variables (e.g., two multiple-choice items, each item being represented in the form of the subjects-by-options (1,0) incidence matrix). Its important link to the topic of our interest lies in its connection to the *bivariate normal distribution*. In other words, dual scaling, which does not employ any assumption about the distribution of responses, has the effect of shaping the two-way data into the form of the best possible multinomial normal distribution by manipulating the spacing of rows and columns. The importance of this aspect of dual scaling will be made clear when we look at polychoric correlation later.

When Kendall and Stuart (1961) discussed the above link of the optimization process to the bivariate normal distribution, it was referred to as "canonical correlation" applied to two categorical variables (i.e., categories of each variable are treated as a set of variables), rather than Hoteling's canonical correlation (Hotelling, 1936) between two sets of variables. canonical correlation à la Kendall and Stuart is what people refer to as canonical correlation in SEM. Keep in mind that this correlation is determined for a pair of categorical variables through appropriate spacings of categories of each variable. Since the category weights of a single variable are determined each time it is paired with another variable, this process is referred to as *pair-wise quantification*.

4. More than Two Categorized Variables

When we have more than two categorized variables, the data are arranged in the form of the subjects-by-"options of the variables" table of the so-called (1,0) incidence matrix. To make this clear, let us indicate by $\mathbf{F}$ the subject-by-category (1,0) incidence matrix of n multiple-choice items,

$$\mathbf{F} = (\mathbf{F}_1, \mathbf{F}_2, \ldots, \mathbf{F}_j, \ldots, \mathbf{F}_n), \tag{4.1}$$

where $\mathbf{F}_j\mathbf{1} = \mathbf{1}$, that is, each subject chooses one category per item. The elements of $\mathbf{F}$ is decomposed into a bilinear form as mentioned earlier. This means that for each component k, we have a single weight vector for rows, $\mathbf{y}_k$, and a single weight vector for columns, $\mathbf{x}_k$. In other words, for the given component, each variable has a fixed vector of category weights for calculation of inter-variable correlation. Thus, dual scaling with more than two variables is referred to as *global quantification*. This is quite different from the two variable case, in which the weight vector of one variable is determined only in relation to the variable paired with it, meaning that for the set of n categorical variables a single variable will end up with $n-1$ weight vectors under pair-wise quantification.

With the new definition of data matrix $\mathbf{F}$ and the corresponding definition of other terms, the formulas (3.7) to (3.12) still hold for dual scaling of multiple-choice data. Let us indicate by n the number of multiple-choice items and by m_j the number of response options (alternatives) of item j. Then, the sum of the squared singular values, excluding 1, is given by

$$\sum_{k=1}^{K} \rho_j^2 = \frac{\sum_{j=1}^{n} m_j}{n} - 1 = \overline{m} - 1. \tag{4.2}$$

If we were to stretch our imagination over the Kendall-Stuart statement to the dual scaling with many categorical variables, we may be able to state that

> "dual scaling of n categorized variables can be viewed as an attempt to determine the spacings of categories of each of n sets in such a way that the resultant n-dimensional contingency table resembles that of a categorized n-variate normal distribution."

This implicit manipulation of the shape of the distribution will later be discussed as an important interpretation of dual scaling when we look into canonical correlation and polychoric correlation. At the present moment, we leave out the distributional aspect of dual scaling, and concentrate on the problem associated with the Young-Householder theorem.

When the $n \times n$ correlation matrix is calculated from variates optimally scaled by dual scaling, it is always positive definite or positive semi-definite. It is so because we have a single matrix of scaled data, say $\mathbf{G}$, and the correlation matrix is obtained as the cross product of $\mathbf{G}$, which guarantees that the correlation matrix is positive definite or positive semi-definite. As such, dual scaling derives numerals that can be mapped in Euclidean space.

5. Pair-wise versus Global Quantification

To compare the two modes of quantification, let us first digress from this problem and discuss some preliminaries of the basics.

5.1 Linear Combination of Variables and Correlation

For the time being, let us assume that a matrix of real numbers is given as data, say $\mathbf{X}$. To make our discussion simple, consider two variables such as scores on two achievement tests so that $\mathbf{X} = (\mathbf{x_1}, \mathbf{x_2})$ and their linear combination $\mathbf{y} = w_1\mathbf{x_1} + w_2\mathbf{x_2}$. It is well known that when the two weights satisfy the relation that $w_1^2 + w_2^2 = 1$ all the composite scores y_i are projections of the data points (x_{1i}, x_{2i}) on the line passing through the origin with the slope of w_1/w_2, assuming that the horizontal axis is X_1 and the vertical axis X_2. If the composite axis goes through the origin $(0,0)$ and Point (a,b), therefore, the projection of the data points onto this composite axis can be expressed as

$$y_i = \frac{a}{\sqrt{a^2 + b^2}} X_{1i} + \frac{b}{\sqrt{a^2 + b^2}} X_{2i} = \cos\theta X_{1i} + \sin\theta X_{2i} = w_1 X_{1i} + w_2 X_{2i}$$

$$(5.1)$$

where the angle θ is the angle between the composite axis and the axis X_1. The important aspect of this expression is that *all the linear combinations are geometrically projections* of data points *on a single axis*.

Let us extend this discussion to n variables, that is, $\mathbf{X} = (\mathbf{x_1}, \mathbf{x_2}, \ldots, \mathbf{x_n})$ and $\mathbf{x_k}$ is $N \times 1$. Consider n orthogonal linear composites, say $\mathbf{Y} = (\mathbf{y_1}, \mathbf{y_2}, \ldots, \mathbf{y_n})$ in the principal hyper-space. Then, this becomes principal component analysis, where for n variables we obtain n orthogonal composites $\mathbf{Y}$, which is $N \times n$, and n vectors of weights $\mathbf{W}$, which is $n \times n$, such that

$$\mathbf{Y} = \mathbf{XW}, \qquad \mathbf{X} = \mathbf{YW'}, \tag{5.2}$$

and the variance of each composite is equal to an eigenvalue. The last expression indicates that the original data matrix of n variables can also be expressed as linear combinations of n orthogonal components. Recall that all linear combinations are projections on straight lines. Thus *scores of a variable* are all *projections on an axis*, going through the origin and point $(w_{1j}, w_{2j}, \ldots, w_{nj})$, that is, a point in n dimensional hyperspace. Notice that this is a proper and logical geometric representation of data in multidimensional space. This provides a stark contrast to the usual practice of representing data, for example, in terms of the correlated axes of mathematics test against English test as if the two axes were orthogonal.

In principal component analysis, we often list the loadings of a variable on all possible components and note that the sum of the squared loadings of each variable is one, that is, the same condition imposed on the weights for a linear combination of variables. Similarly, we know that the sum of the squared loadings of each variable in factor analysis is called the communality with the upper bound of 1. These tables of loadings show that each variable can be expressed as a linear combination of the components or factors and as such scores of subjects on each variable are all aligned along a single axis.

It is well known in factor analysis and principal component analysis that the product moment correlation between two variables can be expressed as the cosine of the angle between the two axes of the variables in this multidimensional Cartesian coordinate system. Such a statement as this presupposes that each variable can be expressed as an axis, rather than what one may think as clouds in the scatter plot of two variables.

Let us now get back to the main discussion. When we subject a pair of categorical variables 1 and 2 to dual scaling the product-moment correlation r_{12} is maximized through weighting of the categories of variables 1 and 2. If we subject variables 1 and 3 to dual scaling the correlation r_{13} is maximized by determining another set of category weights for variable 1 and a new set of category weights for variable 3. If we analyze the data pair-wise, therefore, the category weights of variable 1 used for r_{12} would generally be different from those of variable 1 used for maximizing r_{13}. Thus, this process of pair-wise quantification cannot provide a single matrix of weighted data for the given data. This means that under the pair-wise quantification we cannot define the correlation in the hyper-space as the angle of two data axes, the topic we have just seen above. More simply, if there does not exist a single data matrix, one can no longer calculate the product-moment correlation. The collection of pair-wise correlation coefficients is no longer a set of product-moment correlations, but a set of numbers obtained using different axes each time one variable is paired with another variable. The consequence of pair-wise quantification is empirically known to lead to negative eigenvalues of the correlation matrix. An eigenvalue is nothing but the variance of a composite of n variables, and it is hard to interpret the negative value since the variance is a statistic calculated from the square of measurements. As mentioned earlier, if the triangular inequality is not satisfied by the data, the projection of data on the mean subspace could be longer than the original vector of observation, leading to a negative value of the variance estimate. The absence of a single weighted data matrix means more simply than the above that the correlation matrix can no longer be calculated as the cross product of the entire matrix, the condition that guarantees positive definiteness or positive semi-definiteness. From the dual scaling point of view, this pair-wise procedure is an example of inappropriate use of scaling. And, this is exactly the way in which the Kendall-Stuart canonical correlation is calculated in SEM.

In contrast, *global quantification* determines only a single set of category weights for one variable. Thus, there exists a matrix of quantified data, and the correlation matrix or the variance-covariance matrix is always positive definite or semi-definite. This is the case of dual scaling of multiple-choice data, sometimes called multiple correspondence analysis. That the categories of a single variable have only one set of weights means the resultant correlation matrix under global quantification can be vastly different from that of pair-wise quantification, and on the average, the correlation from global quantification would be considerably lower than that of pair-wise quantification. This higher average correlation under pair-wise quantification, however, does neither guarantee the positive definiteness or semi-definiteness, nor can we find the logical definition in Euclidean hyper space.

Let $\mathbf{x_j}$ be an $m_j \times 1$ vector of weights for m_j categories of item j, $j = 1, 2, \ldots, n$, and $\mathbf{x}$ be an $m \times 1$ vector of weights for all categories of n items, that is, $m = \sum m_j$. Let $\mathbf{D_j}$ be the diagonal matrix of column totals of $\mathbf{F_j}$ and $\mathbf{D_c}$ be that of $\mathbf{F}$. Then, pair-wise quantification determines $\mathbf{x_j}$ and $\mathbf{x_k}$ for items j and k so as to maximize the correlation between the two items. The square of this correlation between two weighted score vectors, $\mathbf{F_j x_j}$ and $\mathbf{F_k x_k}$, is given by

$$r_{jk}^2 = \frac{\mathbf{x_j' F_j' F_k x_k}}{n \mathbf{x_j D_j x_j x_k' D_k x_k}}. \tag{5.3}$$

Under pair-wise quantification, category weights are determined in such a way that $n(n-1)/2$ pairs of correlation coefficients are independently maximized, thus resulting in $n-1$ sets of categories weights for each item. In contrast, global quantification determines a single set of category weights for each item. If we set the origin of scaled measurement to zero (i.e., $\mathbf{1'Fx} = 0$), global dual scaling determines $\mathbf{x}$ for the entire set of items so as to maximize the correlation ratio or squared singular value ρ^2:

Table 1: Sample Data from 50 Subjects on 10 Multiple-Choice
Questions (only chosen option numbers are listed)

1112111112	1112221311	1141112333	1114321114	1111222312
1114312212	2313233322	2124242523	2121111211	1114251212
1111112111	2111212434	1111111111	1111111312	2122121212
3222342221	1111111213	1113111333	1111111411	1111211111
4144441444	1111111111	1111242113	3133323312	2313233322
1112211211	1211211212	1111311424	3143212444	3222342221
2122122211	1111311312	1111223424	1111311424	2313233322
1114251212	2111111131	2131111411	1111111212	1141311333
2313233322	1213211412	2121411244	1111111212	2124242523
1111111212	1144512454	1111311112	1111122111	4122321323

$$\rho^2 = \frac{\mathbf{x'F'Fx}}{n\mathbf{x'D_c x}} = \frac{\sum r_{jt}^2}{n}, \tag{5.4}$$

where r_{jt} is called the item-total correlation and it is the correlation between $\mathbf{F_j x_j}$ and $\mathbf{Fx}$.

Global dual scaling has a number of optimal properties. Among others, it yields maximally reliable scores for the subjects (Lord, 1958) because it maximizes the Cronbach generalized internal consistency reliability, typically called "Cronbach's alpha," or simply indicated as α:

$$\alpha = 1 - \frac{1 - \rho^2}{(n - 1)\rho^2}. \tag{5.5}$$

5.2 Examples of Canonical Correlation and Dual Scaling

An artificial data set of ten categorized ordinal variables from fifty subjects was used for the numerical illustration of the problems. For the benefit of other researchers in the future, we list the data set as seen in Table 1.

Using this data set, we first looked at pair-wise quantification of the Kendall-Stuart canonical correlation (Table 2). This is an option used in SEM as "canonical correlation." As expected, there are negative eigenvalues, meaning that the decomposition of this correlation

Table 2: Canonical correlation : Pair-wise Quantificaiton

1.00									
0.54	1.00								
0.72	0.27	1.00							
0.53	0.69	0.55	1.00						
0.54	0.42	0.54	0.52	1.00					
0.58	1.00	0.56	0.71	0.55	1.00				
0.52	0.80	0.45	0.68	0.45	0.84	1.00			
0.40	0.52	0.65	0.59	0.41	0.63	0.50	1.00		
0.65	0.57	0.77	0.42	1.00	0.71	0.55	0.57	1.00	
0.29	0.40	0.51	0.49	0.62	0.54	0.44	0.73	0.70	1.00

eigenvalues: 6.25, 1.29, 0.88, 0.73, 0.42, 0.27, 0.21, 0.13, −0.04, −0.13

matrix will not provide coordinates for the ten variables in Euclidean space. The same data set was subjected to global quantification of dual scaling, and the 10×10 correlation matrix, based on the first dual scaling solution was obtained (Table 3). All the eigenvalues are positive, hence the correlation matrix is positive definite. The data can be mapped in 10-dimensional Euclidean space. For the sake of future discussion, we have also carried out dual scaling with the weak order constraint (i.e., the equality of two adjacent categories is included, in addition to inequality) on the categories of each variable, using the successive data modification method (Nishisato, 1980). The correlation matrix obtained from the first solution under the order constraint is listed in Table 4. Like the case in Table 3, this correlation matrix contains only positive eigenvalues.

As can be seen in the three tables, global quantification, with or without order constraints on the categories of variables, always yields positive definite or semi-definite matrix, that is, no negative eigenvalues. In contrast, pair-wise quantification has produced two negative eigenvalues, and this indicates the failure of the requirement that variables lie in Euclidean space. At the same time, the average correlation under pair-wise correlation is much higher than that of the two procedures of global quantification. Note, however, that such high correlation is hardly interpretable, of little use, or even worse, misleading and nonsensical.

Table 3: Correlation by dual scaling : global quantification

1.00									
0.46	1.00								
0.16	0.08	1.00							
0.36	0.68	0.05	1.00						
0.15	0.39	0.12	0.37	1.00					
0.50	0.97	0.09	0.66	0.42	1.00				
0.41	0.79	0.20	0.66	0.39	0.82	1.00			
0.19	0.45	−.07	0.48	0.27	0.49	0.46	1.00		
0.40	0.54	0.12	0.30	0.39	0.58	0.52	0.35	1.00	
−.06	0.36	0.17	0.34	0.24	0.33	0.31	0.36	0.01	1.00

eigenvalues: 4.68, 1.23, 1.08, 0.83, 0.63, 0.55, 0.46, 0.29, 0.22, 0.02

Table 4: Correlation from Global dual scaling with order constraints

1.00									
0.39	1.00								
0.16	−.14	1.00							
0.31	0.42	0.19	1.00						
0.16	0.26	0.06	0.44	1.00					
0.46	0.54	−.10	0.55	0.44	1.00				
0.41	0.73	0.08	0.43	0.32	0.49	1.00			
0.26	0.27	0.39	0.34	0.35	0.14	0.36	1.00		
0.50	0.35	0.24	0.29	0.38	0.35	0.39	0.52	1.00	
−.06	0.16	0.12	0.38	0.37	0.13	0.19	0.40	0.26	1.00

eigenvalues: 3.92, 1.56, 1.19, 0.80, 0.54, 0.40, 0.34, 0.30, 0.23

6. Polychoric Correlation

Polychoric correlation is based on the assumption of the underlying multivariate normal distribution. Thus, this is quite different from canonical correlation and dual scaling where no distributional assumptions are used in the process of deriving the results. In contrast, polychoric correlation is calculated using the joint normal distribution. Let us start with a brief introduction to polychoric correlation.

6.1 Background
When polychoric correlation is used for two categorized ordinal variables, one must determine the thresholds (category boundaries) for

each variable, assuming that the two variables are latently distributed as bivariate normal. It is therefore natural to determine the thresholds to make the bivariate multinomial distribution approximate that of categorized bivariate normal distribution.

When there are only two categorized ordinal variables, polychoric correlation was extensively studied with the 3×3 contingency tables by Tallis (1962). When the dimension of the categorized table is extended to $r \times s$, Martinson and Hamdan (1971) and Olsson (1979) developed methods for estimating polychoric correlation. Poon and Lee (1987) mention Lee's attempt to generalize it to $r \times s \times t$ table (i.e., three categorized ordinal variables), in their paper in which they extended the case from bivariate to multivariate normal distribution. Lee, Poon and Bentler (1992) discuss the estimation problem associated with the structural equation models with polychoric correlation, assuming the multivariate normal distribution. In their paper, they point out that Muthén's estimates (1984) are not necessarily based on the multivariate normal distribution. Lee, Poon and Bentler (1992) then propose a generalization of the results in Lee, Poon and Bentler (1990a, 1990b), to continuous and polytomous variables, which is also a generalization of Olsson, Drasgow and Doran (1982) and Poon and Lee (1987). These are only a few frequently cited papers, and there must be many other relevant papers.

6.2 Threshold Determination: Pair-Wise or Univariate

From the above sketchy review of literature it seems clear that at least the knowledge exists on the estimation of polychoric correlation based on the multivariate normal distribution. Whether the knowledge is too difficult to put into practical use or not, it is an unexpected discovery that some of the widely used computer programs for structural equation modeling (SEM) seem to employ either pairwise determination of thresholds, rather than global threshold determination (Hemsworth, 1999), or univariate determination of thresholds (Nishisato, 2000). In the former of pair-wise determination of thresholds, for n variables, the categories of each variable take in general n-1 different sets of thresholds (pair-wise operation), rather than one set (global operation). This is very much like pair-wise quantification of the Kendall-Stuart canonical correlation, except that polychoric correlation between two variables is calculated using the bivariate normal distribution. In the latter case of univariate determination of thresholds, each variable is given a single set of thresholds, calculated from the category frequencies of each variable molded under the univariate normal distribution, and this is done independently of the other $n - 1$ variables. But it is well known in statistics that a set of n univariate normal distributions does not guarantee that they are jointly multivariate normal.

A surprising aspect of the matter is that the problem of non-positive definiteness or non-positive semi-definiteness is not restricted to PRELIS (Jöreskog and Sorbom, 1996), but can also be observed in such programs as CALIS (SAS Institute, 1991), EzPath (Steiger, 1989), EQS (Bentler, 1989) and LISCOMP (Muthén, 1988). Furthermore, this problem is not limited to this area of study, but also extended to other areas of multivariate analysis, such as exploratory factor analysis (MacCallum, 1983). What are the consequences of these two procedures? The case of pair-wise quantification problems has already been discussed. We should therefore direct our attention to the univariate determination of thresholds.

6.3 Multivariate Normality

It is important to discuss in more details how polychoric correlation is calculated. The difficulty in using the univariate method of threshold determination seems to lie in the fact that polychoric correlation is calculated by forcing the observed data into the framework of the latent multivariate normal distribution.

To look at this molding of the data into multivariate normality, let us examine one of the popular formulas for calculating polychoric correlation. For a $c \times d$ table of two ordinal variables, x and y, Olsson (1979) and Bollen (1989) provide the formula to calculate the log likelihood ratio as,

$$\ln L = A + \Sigma_i^c \Sigma_j^d N_{ij} \ln(\pi_{ij}), \qquad (6.1)$$

where c and d are the numbers of categories for the two variables, A a constant, and N_{ij} is the response frequency in the cell (i, j) of the contingency table. The thresholds for variable X are a_i, $i = 0, 1, 2, \ldots, c$ and those for Y are b_j, $j = 0, 1, 2, \ldots, d$, where a_0 and b_0 are $-\infty$, and

a_c and b_d are $+\infty$. Note also that

$$\pi_{ij} = \Phi_2(a_i, b_j) - \Phi_2(a_{i-1}, b_j) - \Phi_2(a_i, b_{j-1}) + \Phi_2(a_{i-1}, b_{j-1}), \quad (6.2)$$

where $\Phi(\cdot, \cdot)$ is the bivariate normal distribution function with correlation ρ.

Thus, the formula indicates that observed data are forced into, or regressed to, the bivariate normal distribution, and this must be the case for all possible combinations of variables. This means that the set of n variables must follow the multivariate normal distribution, and that this is the assumption on which polychoric correlation is calculated. It is well known that if a set of variables follows the multivariate normal distribution any pair of these variables has a bivariate normal distribution. The converse, however, is not true: there is no guarantee that a collection of univariate normal distributions is jointly multivariate normal. The univariate determination of thresholds seems to be only an attempt to substitute a collection of univariate normal distributions for the joint multivariate normal distribution. An immediate question is what will happen if the underlying distribution is not multivariate normal and yet we use the univariate determination of thresholds. We seem to be forcing or regressing non-normal variates to normal variates, leading to a very poor regression result as the latent distribution of observed data moves away from the multivariate normal distribution.

Thus it is our interpretation that the problem of non-positive definiteness or semi-definiteness is not solely caused by the methods of threshold determination, but also by the departure from the latent multivariate normal distribution. When the underlying distribution is far from the multivariate normal distribution, polychoric correlation loses its theoretical foundation.

6.4 Numerical Example of Polychoric Correlation

The same data set as used in the previous section (i.e., Table 1) was used to calculate a 10×10 matrix of polychoric correlation by PRELIS (Table 4).

Table 5: Matrix of polychoric correlation by PRELIS

```
1.00
0.44   1.00
0.70  -.31   1.00
0.43   0.37   0.34   1.00
0.35   0.23   0.40   0.48   1.00
0.50   0.46   0.13   0.72   0.30   1.00
0.43   0.74   0.19   0.35   0.19   0.49   1.00
0.36   0.15   0.48   0.39   0.32   0.15   0.29   1.00
0.55   0.23   0.69   0.33   0.59   0.06   0.36   0.55   1.00
0.08  -.25   0.30   0.29   0.58   0.02   0.11   0.54   0.69   1.00
```
eigenvalues: 4.26, 2.17, 1.07, 0.98, 0.68, 0.43, 0.26, 0.25, 0.08, -0.17

The result shows one negative eigenvalue. Should we ignore this single negative eigenvalue? If we do, it is very much like ignoring the case in which an estimate of a probability measure becomes either negative or greater than 1. Then, it is no longer a probability measure. The same reasoning should at least be considered here, because some variables may no longer be represented in Euclidean space, and furthermore as a consequence of the above, the relations between variables may no longer be accurately captured by the correlation measure. One of the general attitudes to this kind of problem is that at least the first few components have positive eigenvalues, hence we may adopt only those "major" components. But, where can we find the rationale that we may get away with the logical problem by ignoring higher dimensional space? There does not appear to be any guarantee that the low-dimensional representation of data is accurate enough to carry out detailed analysis of data structure.

As mentioned earlier, let us remember that if n variables follow the multivariate normal distribution any pair of variables follows the bivariate normal distribution, but the converse does not generally follow. In other words, a collection of univariate normal distributions does not guarantee that they are jointly multivariate normal. Then it seems to suggest that one cannot use the univariate method of threshold determination for polychoric correlation as used in SEM.

Table 6: Subscales, Numer of Items and Number of Subjects

Subscale	No.of Items	No. of Subjects
Anxiety	10	116
Depression	13	106
Hostility	6	123
Interpersonal Sensitivity	9	116
Obsessive Compulsive	10	117
Paranoid Ideation	6	119
Phobia	7	115
Psychoticism	10	111
Somatization	12	112

7. Further Empirical Findings

The above numerical examples may not convince us that matrices of canonical correlation and polychoric correlation often fail to be positive definite or semi-definite. But, so far as the literature can tell, such failures are rampant in practice. Therefore, we would like to present further results reported in Hemaworth (2002) which are based on the analysis of real data.

The data collected by Susan James are a standardized clinical psychology battery (SCL90, unpublished) with 90 five-point Likert scale questions, divided into nine subscales, anxiety, depression, hostility, interpersonal sensitivity, obsessive compulsive, paranoid ideation, phobia, psychoticism, and somatization. Each subscale is intended to have one latent variable with a varying number of measurement variables (questions). Although 133 respondents completed the SCL90, missing values were deleted list-wise. Thus, at the final stage, the number of respondents retained in the data set and that of questions of each subscale are, respectively, as in Table 6.

For each subscale, 50 bootstrapped samples were drawn with replacement from the parent subscale, using PRELIS (Jöreskog and Sor bom, 1996). A seed value of 1234567 replaced the random seed so that an identical set of bootstrapped samples would be used for each of the three methods, the Kendall-Stuart canonical correlation, polychoric correlation and dual scaling. Table 7 shows the number of positive (semi-)definite matrices, out of 50, for each subscale under the three methods.

Table 7: Number of positive definite, Semi-Definite Matrices out of 50

Subscale	Canonical	Polychoric	Dual
Anxiety	1	34	50
Depression	0	8	50
Hostility	25	45	50
Interpersonal Sensitivity	10	43	50
Obsessive Compulsive	2	44	50
Paranoid Ideation	38	50	50
Phobia	23	45	50
Psychoticism	1	15	50
Somatization	1	34	50

Note that all the matrices under dual scaling are positive definite or semi-definite. Unlike the previous results from an artificial data set, the present results clearly show that the Kendall-Stuart canonical correlation is inappropriate for use in SEM or anything else. polychoric correlation also produces a number of inappropriate correlation matrices. Since we are dealing with ordered categorical variables, it is possible to use the product-moment correlation of Likert-scaled data, that is, by assigning, for example, 1, 2, 3, 4 and 5 for the five categories of a variable. The correlation matrix of product-moment correlation is always positive definite or semi-definite. The difference between the case of product-moment correlation and dual scaling is that dual scaling quantifies the data so as to maximize correlation, and that in this process dual scaling molds the distribution as close as possible to the categorized bivariate normal distribution. Thus, the average of correlation coefficients from the product-moment correlation of Likert-scaled variables is always lower than, or equal to, that from dual scaling.

In the results on the Kendall-Stuart canonical correlation in Table 7, we note a relation between the failure of producing acceptable correlation matrices and the number of variables (i.e., items). The least-squares regression equation (Hemsworth, 2002) is

$$y = -4.4x + 76, \tag{7.1}$$

where y is the number of positive definite or semi-definite matrices and

x is the number of variables, and the regression coefficient is significant. Thus, the number of positive definite or semi-definite matrices
decreases as the number of measurement variables increases.

8. What Alternatives if Any?

We have identified two possible causes of the problem that canonical correlation and polychoric correlation as used in SEM encounter,
that is, negative eigenvalues. They are the methods of threshold determination, pair-wise or univariate, and the lack of the multivariate
normality of the latent joint distribution. Researchers in the area of
SEM have overcome this problem of negative eigenvalues by devising
practical approaches to it: Smooth out the offending indefinite matrices by (a) adding a ridge (implemented as the default in LISREL) or
(b) using only positive principal components. The latter is referred
to as principal component smoothing, and this has been used to overcome negative eigenvalues that are routinely observed with tetracholic
correlation matrices. The method is used in the following way,

(1) Determine eigenvalues and eigenvectors of the matrix,

(2) Using only positive eigenvalues and associated eigenvectors, calculate the semidefinite components of the correlation matrix. The
resulting matrix will have larger diagonal entries on average than the
original matrix.

(3) If all variances in the diagonal of the original matrix were strictly
greater than zero, rescale the reproduced matrix to these original variances.

This method is already incorporated in Testfact program (Wilson,
Wood and Gibbons, 1984). Our question is on how to justify this
kind of procedure or on what kind of assurance we can have about
the validity of the final outcome. This problem of validity is of great
importance for data analysis.

The fact that the original correlation matrix contains at least one
negative eigenvalue indicates that those elements in the correlation

matrix are all very questionable estimates of the correlation coefficients. This doubt cannot be washed off by just ignoring negative eigenvalues, but it is a more fundamental problem than that. It is not just with negative eigenvalues but with the entire outcome related to the decomposition that we must be concerned with. To be extremely critical, it would be like a problem that the trace of the 10×10 correlation matrix is 10 as we all know, but that the eigenvalues obtained from the matrix were, for example, 10, 5, 3, 2, 1, 0, −1, −2, −3, and −5. In the above approach, we will use the first five components, one of which has the eigenvalue of 10, that is, 100 per cent of the sum of all the possible eigenvalues! With this example, we can see how absurd the above procedure can get in an extreme case. It seems clear from the numerical examples and discussion that dual scaling offers a very promising alternative. It is based on global quantification, and as we suspect it approximates multivariate normality in the course of maximizing the average pair-wise correlation of the entire data set. In the previous section, it was briefly mentioned that the product-moment correlation matrix of Likert-scaled variables also guarantees its positive definiteness or semi-definiteness. However, the problem of Likert-scaling is that it cannot capture nonlinear relations among variables, nor provides any means of normalizing of skewed distributions, not to mention the shape of the joint distribution. Thus, it appears clear that dual scaling is the winner among those methods discussed in this paper.

Some caution, however, must be exercised here. Before we decide to replace the Kendall-Stuart canonical correlation and polychoric correlation in SEM with dual scaling, one must investigate the behavior of dual scaling as a function of the number of categories (alternatives, options) of each variable. This is so because pair-wise correlation under dual scaling is likely to increase monotonically as the number of categories of each variable increases and because we do not know what effects the shape of the underlying (latent) distribution might affect this monotone increase. Thus, the phrase "the higher the better" does not apply to the current problem. We must investigate first the following questions:

"What is the optimal number of categories of each variable in order to estimate the underlying correlation?"

"Does the average product-moment correlation from dual scaling monotonically increases as the number of categories increases, even when the latent variables are multivariate normal?"

These are difficult questions to answer, for the first question requires a different kind of criterion for optimality, rather than the traditional "maximal correlation". Because if we use the correlation as the criterion, we would conclude that the more categories the better. This conclusion, which we do not think is reasonable, comes from the fact that the total information, contained in item j, is defined as the sum of squared item-total correlations in the entire space and it is directly related to the number of categories of each item (Nishisato, 1994, 1996),

$$\sum_{k=1}^{K} r_{jt}^2 = m_j, \tag{8.1}$$

where m_j is the number of categories of variable j. For an alternative criterion for optimality, there is some indication of possibility. Nishisato (1998) indicates that the Cramér coefficient and the Tchuprov coefficient of an $n \times m$ contingency are a concave function of the numbers of categories, n and m. In other words, these coefficients between two categorical variables tend to increase when the number of categories (options) increases to a certain point and then declines. But, this observation is currently under scrutiny and we must wait for the results of the investigation.

9. Concluding Remarks

The main purpose of the current paper was to draw the attention of the researchers to the importance of the role of scaling when data are categorical. In particular, when we have multivariate data, a typical approach to data analysis involves the construction of composite variates, the most efficient being those of principal components. Dual scaling is known to transform data linearly or nonlinearly in such a way that the variance of the composite of the transformed data attains its maximum. It is known (e.g., Lord, 1958; Nishisato and Sheu, 1980) that such a composite of a maximal variance has the property to maximize the average correlation between variables and the composite. In this context, the statement by Kendall and Stuart (1961) becomes relevant to the current paper: that maximization of the correlation by adjusting the spacing of categories of each variable is in essence the same operation to space categories of the corresponding contingency table such that the resultant contingency table with such spacing of categories resembles that of a partitioned bivariate normal distribution. In the context of global quantification of dual scaling the Kendall-Stuart statement offers a support to the operation such that global dual scaling which transforms n categorical variates to maximize the average inter-variable correlation is in essence the same as transforming the n-dimensional contingency table into the n-dimensional multinomial distribution that resembles that of the partitioned n-variate normal distribution. As such, dual scaling finds justifiable room for its contribution to the study of structural equation modeling (SEM).

From the user's point of view, it is reasonable to ask the question on how we should interpret the results based on the Kendall-Stuart canonical correlation or polychoric correlation from pair-wise thresholds or univariate thresholds. A convincing answer cannot be found anywhere because there does not seem to be any. In the current study, we identified not only the problem of positive definiteness or semi-definiteness but also the possible lack of multivariate normality of underlying variables.

Let us conclude this paper by making a bold conjecture. Suppose that the underlying multivariate distribution is indeed normal. What should we expect from categorized ordinal variables from such a distribution? Our bold conjecture is as follows: the pair-wise quantification and the global quantification should provide the identical results, and the Kendall-Stuart canonical correlation, polychoric correlation and dual scaling must all converge to the same sample correlation matrix. Assuming that this conjecture is correct, what seems to be important is to examine if the assumption of underlying multivariate normality is appropriate for a given data set. If it is, then use polychoric correlation with the univariate method of threshold determination; if it is not, use dual scaling. No matter what underlying multivariate distribution we may have, it seems that dual scaling will offer a justifiable means to calculate the correlation. We should be reminded, however, the quest for a better statistic should not end here. What about the information captured by the second and subsequent solutions of dual scaling ?

Acknowledgements

The work was partially supported by a grant from the Natural Sciences and Engineering Council of Canada to S. Nishisato.

References

Beltrami, E. (1873). Sulle funzioni bilineari (On the linear functions). In G. Battagline and E. Fergola (Eds.), *Giornale di Mathematiche*, **11**, 98-106.

Bentler, P. M. (1989). *EQS structural equations program manual.* Los Angeles: BMDP Statistical Software.

Benzécri, J.-P. and Cazes, P. (1973). *L'Analyse des Données: II. L'Analyse des Correspondances (Data analysis II: Correspondence analysis.* Paris: Dunod.

Bock, R. D. (1960). Methods and applications of optimal scaling. *The University of North Carolina Psychometric Laboratory Research Memorandum*, No.5.

Bollen, K. A. (1989). *Structural Equations with Latent Variables.* New York: Wiley-Interscience.

Bradley, R. A. and Terry, M. E. (1952). Rank analysis ofincomplete block design: The method of paired comparisons. *Biometrika*, **39**, 324-345.

Coombs, C. H. (1964). *A theory of data.* New York: Wiley.

Cronbach, L. J. (1951). Coefficient alpha and the internal structure of tests. *Psychometrika*, **16**, 297-334.

Escofier-Cordier, B. (1969). L'analyse factorielle des correspondances. *Bureau Universitaire de Recherche Operationelle, Cahiers, Série Recherche*, **13**, 25-29.

Gifi, A. (1990). *Nonlinear multivariate analysis.* New York: Wiley.

Greenacre, M. J. (1984). *Theory and applications of correspondence Analysis.* London: Academic Press.

Gorsuch, R. L. (1983). *Factor Analysis.* 2nd ed. Hillsdale, N.J.: Lawrence Erlbaum.

Guttman, L. (1950). Chapters 3-6. *Measurement and prediction*, edited by Stouffer, S. A., et al., Princeton: Princeton University Press.

Hamming, R. W. (1950). Error detecting and error correcting codes. *The Bell System Technical Journal*, **26**, 147-160.

Hayashi, C. (1950). On the quantification ofqualitative data from the mathematico-statistical point of view. *Annals of the Institute of Statistical Mathematics*, **2**, 35-47.

Hayashi, C. (1952). On the prediction of phenomenon from qualitative data and the quantification of qualitative data from the mathematico-statistical point of view. *Annals of the Institute of Statistical Mathematics*, **3**, 69-98.

Hemsworth, D. (1999). Personal communication with Stephen DuToit.

Hemsworth, D. (2002). The use of dual scaling for the production of correlation matrices for use in structural equation modeling. Unpublished Ph.D. thesis, University of Toronto.

Hill, M. O. (1974). Correspondence analysis: A neglected multivariate method. *Applied Statistics*, **23**, 340-354.

Hirshcfeld, H. O. (1935). A connection between correlation and contingency. *Cambridge Philosophical Society Proceedings*, **31**, 520-524.

Horst, P. (1935). Measuring complex attitudes. *Journal of Social Psychology*, **6**, 369-374.

Hotelling, H. (1936). Relation between two sets of variables. *Biometrika*, **28**, 321-377.

Jordan, C. (1874). Mémoire sur les formes bilineares (Notes on bilinear forms). *Journal de Mathématiques Pures et Appliquées, Deuxiéme Série*, **19**, 35-54.

Jöreskog, K. and Sorbom, D. (1996). *Lisrel 8: Users reference guide.* Chicago: Scientific Software International.

Jöreskog, K. and Sorbom, D. (1996). *Prelis 2: Users reference guide.* Chicago: Scientific Software International.

Kendall, M. G. and Stuart, A. (1961). *The Advanced Theory of Statistics.* Volume II. London: Griffin.

Kruskal, J. B. (1964a). Multidimensional scaling by optimizing goodness of fit to a nonmetric hypothesis. *Psychometrika*, **29**, 1-28.

Kruskal, J. B. (1964b). Nonmetric multidimensional scaling: a numerical method. *Psychometrika*, **29**, 115-129.

Lee, S. Y., Poon, W. Y., and Bentler, P. M. (1990a). A three-stage estimation procedure for structural equation models with polytomous variables. *Psychometrika*, **55**, 45-52.

Lee, S. Y., Poon, W. Y., and Bentler, P. M. (1990b). Full maximum likelihood analysis of structural equation models with polytomous variables. *Statistics and Probability Letters*, **9**, 91-97.

Lee, S. Y., Poon, W. Y., and Bentler, P. N. (1992). Structural equation models with continuous and polytomous variables. *Psychometrika*, **57**, 89-105.

Likert, R. (1932). A technique for the measurement of attitudes. *Archives of Psychology*, **140**, 44-53.

Lingoes, J. C. (1964). Simulataneous linear regression: An IBM 7090 porgram for analyzing metric/nonmetric data. *Behavioral Science*, **9**, 87-88.

Lord, F. M. (1958). Some relations between Guttman's principal components of scale analysis and other psychometric theory. *Psychometrika*, **23**, 291-296.

Luce, R. D. (1959). *Individual choice behavior*. New York: Wiley.

MacCallum, R. (1983). A comparison of factor analysis program in SPSS, BMDP, and SAS. *Psychometrika*, **48**, 223-231.

Martinson E. O., and Hamdan, M. A. (1971). Maximum likelihood and some other asymptotically efficient estimators of correlation in two-way contingency tables. *Journal of Statistical Computation and Simulation*, **1**, 45-54.

Messick, S. J., and Abelson, R. P. (1956). The additive constant problem in multidimensional scaling. *Psychometrika*, **21**, 1-17.

Minkowski, H. (1896). *Geometrie der Zahlen*. Leipzig: Teubner.

Muthén, B. (1984). A general structural equation model with dichotomous, ordered categorical and continuous latent variable indicators. *Psychometrika*, **49**, 115-132.

Muthén, B. (1988). LISCOMP: Analysis of linear structural equations with a comprehensive measurement model. Chicago: Scientific Software.

Nishisato, S. (1980). *Analysis of categorical data: dual scaling and its applications*. Toronto: University of Toronto Press.

Nishisato, S. (1993). On quantifying different types of categorical data. *Psy-

chometrika, **58**, 617-629.

Nishisato, S. (1994). *Elements of dual scaling.* Hillsdale, N.J.: Lawrence Erlbaum Associates.

Nishisato, S. (1996). Gleaning in the field of dual scaling. *Psychometrika*, **61**, 559-599.

Nishisato, S. (1998). Unifying a spectrum of data types under a comprehensive frame work for data analysis. A talk presented at a symposium at the Institute of Statistical Mathematics, Tokyo, Japan.

Nishisato, S. (2000). Personal communication with Karl Jöreskog.

Nishisato, S., and Sheu, W. J. (1980). Piecewise method of reciprocal averages for dual scaling of multiple-choice data. *Psychometrika*, **45**, 467-478.

Olsson, U. (1979). Maximum likelihood estimation of the polychoric correlation coefficient. *Psychometrika*, **44**, 485-500.

Olsson, U., Drasgow, F., and Doran, N. J. (1982). The polyserial correlation coefficient. *Psychometrika*, **47**, 337-347.

Poon, W. Y., and Lee, S. Y. (1987). Maximum likelihood estimation of multivariate polyserial and polychoric correlation coefficients. *Psychometrika*, **52**, 409-430.

Richardson, M. and Kuder, G. F. (1933). Making a rating scale that measures. *Personnel Journal*, **12**, 36-40.

Sas Institute Inc. (1991). The CALIS procedure: Analysis of covariance structures. Cary, N.C.

Schmidt, E. (1907). Zür Theorie der linearen und nichtlinearen Integleichungen Erster Teil. Entwicklung willkürlicher Functionen nach Syetemen vorgeschriebener. *Mathematische Annalen*, **63**, 433-476.

Shepard, R. N. (1962a). The analysis of proximities: Multidimensional scaling with an unknown distance function. I. *Psychometrika*, **27**, 125-140.

Shepard, R. N. (1962b). The analysis of proximities: Multidimensional scaling with an unknown distance function. II. *Psychometrika*, **27**, 219-245.

Steiger, J. H. (1989). EzPATH: A supplementary module for SYSTAT and SYGRAPH. Evanston, Il.: SYSTAT.

Stevens, S. S. (1951). *Handbook of Experimental Psychology.* New York: Wiley.

Tallis, G. (1962). The maximum likelihood estimation of correlation from contingency tables. *Biometrics*, **18**, 342-353.

Thurstone, L. L. (1927). A law of comparative judgment. *Psychological Review*, **34**, 273-286.

Torgerson, W. S. (1952). Multidimensional scaling. I. Theory and method. *Psychometrika*, **17**,401-419.

Torgerson, W. S. (1958). *Theory and Methods of Scaling.* New York: Wiley.

Wilson, D., Wood, R., and Gibbons, R. (1984). Testfact: Test scoring, item statistics, and item factor analysis. Chicago: Scientific Software.

Wothke, W. (1993). Nonpositive definite matrices in structural modeling. *Testing structural equation models*, edited by Bollen, K. and Long, S., 256-293, Newbury Park: Sage.

Yanai, H., Shigemasu, K., Maekawa, S., and Ichikawa, M. (1990). *Inshi Bunseki (Factor Analysis).* Tokyo: Asakura Shoten (in Japanese)

Young, G., and Householder, A. S. (1938). A note on multi-dimensional psychophysical analysis. *Psychometrika*, **6**, 331-333.

MTV and MGV:
Two Criteria for Nonlinear PCA

Tatsuo Otsu and Hiroko Matsuo

Department of Behavioral Science
Hokkaido University
N.10 W.7, Kita-ku
Sapporo 060-0810, Japan
otsu@let.hokudai.ac.jp

Summary: MTV (Maximizing Total Variance) and MGV (Minimizing Generalized Variance) are popular criteria for PCA with optimal scaling. They are adopted by the SAS-PRINQUAL procedure and OSMOD (Saito and Otsu, 1988). MTV is an intuitive generalization of linear PCA criterion. We will show some properties of nonlinear PCA with these criteria in an application to the data of NLSY79 (Zagorsky, 1997), a large panel survey in the U.S., conducted over twenty years. We will show the following. (1) The effectiveness of PCA with optimal scaling as a tool for large social research data analysis. We can obtain useful results when it complements analyses by regression models. (2) Features of MTV and MGV, especially their abilities and deficiencies in real data analysis.

1. Introduction

In the social and behavioral sciences, we often need data analysis methods for multivariate data that contain discrete variables. Multiple correspondence analysis (MCA), that is equivalent to Hayashi's quantification method III, is one of the most popular method for this purpose (Lebart et al., 1984; Nishisato, 1994). Another popular method is principal component analysis accompanied by optimal scaling. (Hereafter we abbreviate this as PCAOS.) These methods were categorized under the term "homogeneity analysis" by European psychometricians (Gifi, 1990; Heiser and Meulman, 1995). The PRINQUAL procedure of the SAS system is the most popular implementation of PCAOS. OSMOD (Saito and Otsu, 1988; Otsu and Saito, 1990; Otsu, 1993) is another implementation using the fast optimization algorithm with quasi-Newton projection method. Kuhfeld and Young (1989) showed a criticism against the numerical comparison in Saito and Otsu (1988). One closely related method is variables clustering with optimal scaling (Tsuchiya, 1995), which is a search procedure for oblique factors with optimal scaling.

In MCA, we can obtain the solutions by the one time computations of an eigenvalue problem. On the other hand, we need iterative nonlinear optimization for PCAOS. Although this feature makes the program complicated, we can analyze mixed measurement datasets that contain both discrete and continuous variables. Another advantage of PCAOS is that it is able to impose more restrictions on categories than MCA. This frequently leads to more stable and interpretable solutions. Current theory for PCAOS is not based on rigid mathematics, and its naive correlation estimation between categorical variables often shows under-estimation. In spite of these unsophisticated features, PCAOS can obtain useful insights from real social/behavioral data.

In this article, at first we point out some problems of MCA, then explain the model of PCAOS. The most important feature is the model estimation criterion. Some properties of two criteria, MTV and MGV, are explained. An example of OSMOD analysis on large survey data (NLSY79) is shown.

2. Inadequate Solutions by Multiple Correspondence Analysis

Here we consider some properties of MCA. For a demonstration example, we use the following artificial data.

1. Generate 100 samples of 5-dimensional multivariate normal distribution that has the covariance matrix shown in Table 1.

2. For each variable, categorize their values into 5 categories by rank. The samples of rank 1 to 20 are given category 1, and the samples of rank 21 to 40 are given category 2, and so on.

3. Therefore we obtain 5 discrete variables, each of them having 5 categories. Make 0-1 valued 25 dummy variables from these 5 variables. These 25 dummy variables are analyzed by MCA.

The estimated correlations (singular values) by MCA are shown in Table 2. Configurations of category scores are shown in Figure 1

Table 1: Covariance Matrix of Test Data

Variables	X_1	X_2	X_3	X_4	X_5
X_1	1.0				
X_2	0.7	1.0			
X_3	0.5	0.7	1.0		
X_4	0.2	0.5	0.7	1.0	
X_5	0.1	0.2	0.5	0.7	1.0
Eigenvalues	2.971	1.239	0.400	0.229	0.162
Eigenvectors					
x_1	0.369	0.476	0.525	0.476	0.369
x_2	0.586	0.396	0.000	-0.396	-0.586

and Figure 2. In the figures, symbols such that **1A** show the variables and categories. Numbers show the items (variables) and alphabetical letters show the categories. The first PCA component of the covariance matrix is obtained as the first component of the MCA analysis. But the second component of the MCA solution is the quadratic polynomial of the first component. The second PCA component of the original covariance structure is embedded in the higher order MCA solutions. In this case, the solution of MCA might lead the analyst into the wrong interpretation.

We usually interpret MCA solutions by adopting a few score dimensions. The dimension adoption is based on the values of the correlations (singular values). The adopted dimensions correspond to the largest correlations. In the above example, the adopted two dimensions show inadequate latent data structures. In this example, we know the latent structure. Therefore the inadequacy of the method can be recognized. But in real applications, these insights may be difficult.

Table 2: Estimated Correlations by MCA (Item-Category Type Data)

r_1	r_2	r_3	r_4
0.77	0.60	0.56	0.54

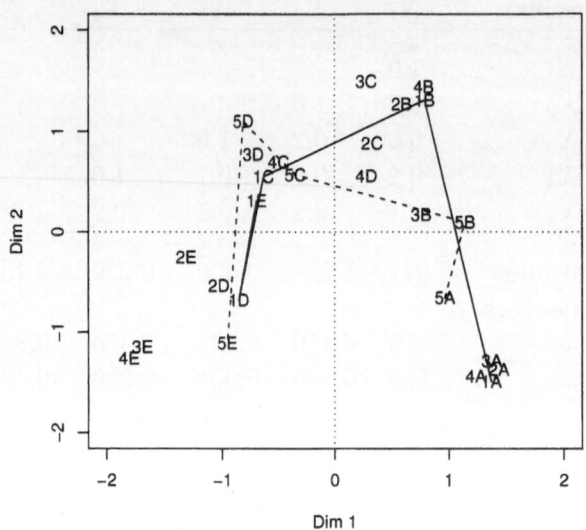

Figure 1: Category Scores by MCA (Dim. 1 vs. Dim. 2)

3. Optimal Scaling with Multivariate Normal Data

The reason for the inadequate solutions is as follows. Some low order polynomials of the first optimal scores, which are orthogonal to the latter, have rather large correlations. These low order polynomials raise solutions that are not intended. Assuming latent multivariate normal structure, these phenomena are clearly explained. In the following, we show some properties of multivariate normal variables and their polynomials. This explanation is based on Bekker and de Leeuw (1988) and Otsu (1993).

Hermite polynomials $\{H_r(x), r = 0, 1, 2, \ldots\}$ play important roles in inspecting nonlinear transformations of normal variables. They are defined as follows.

$$H_r(X) = \frac{(-1)^r}{N(X)} \frac{d^r}{dX^r} N(X) \quad (r = 0, 1, 2, \ldots), \qquad (3.1)$$

where

$$N(X) = \frac{1}{\sqrt{2\pi}} \exp(-X^2/2). \qquad (3.2)$$

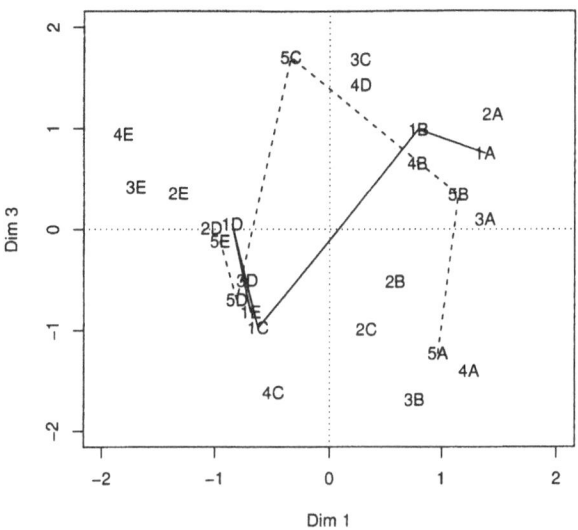

Figure 2: Category Scores by MCA (Dim. 1 vs. Dim. 3)

We can see the detailed properties of Hermite polynomials in Lancaster (1958), Stuart and Ord (1987) and Hushimi and Akai (1981).

Hermite polynomial $H_r(x)$ is a polynomial of order r. The following equations hold.

$$\int_{-\infty}^{+\infty} H_r(X)H_s(X)N(X)dX = 0, \quad (r \neq s)$$

$$\int_{-\infty}^{+\infty} H_r(X)H_r(X)N(X)dX = r! . \tag{3.3}$$

Let $G_r(X) = H_r(X)/\sqrt{r!}$, then $\{G_r(X)|r = 1, 2, \ldots\}$ compose an orthonormal system with respect to the inner product defined by normal weight.

Suppose that X_j and X_k are two-variate normal variables. Also suppose that their variances are ones and their covariance is σ_{jk}. Under these conditions, the following formula holds.

$$E[G_r(X_j)G_s(X_k)] = \delta_{rs}\sigma_{jk}^r , \tag{3.4}$$

where δ_{rs} shows Kronecker's delta.

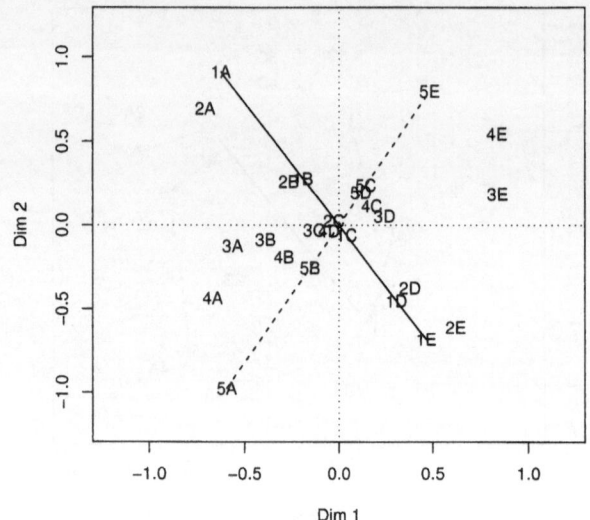

Figure 3: Category Scores by OSMOD with MGV (Dim. 1 vs. Dim. 2)

Suppose a more general polynomial

$$Z_j = g_j(X_j) = \sum_r a_{jr} G_r(X_j) \ . \tag{3.5}$$

The covariance between Z_j and Z_k is given by

$$\mathrm{cov}(Z_j, Z_k) \ = \ \sum_r a_{jr} a_{kr} \sigma_{jk}^r \ . \tag{3.6}$$

We can use the above formulae to analyze the properties of MCA for item-categories. Suppose p-variate normal variables $X_1, X_2, \ldots, X_p$. Let $\boldsymbol{X} = (X_1, \ldots, X_p)'$, and suppose that the distribution of $\boldsymbol{X}$ is $N(\boldsymbol{0}, \boldsymbol{\Sigma})$.

The covariance between the normalized Hermite polynomials is given as follows.

$$E[G_r(X_j) G_s(X_k)] = \delta_{rs} \sigma_{jk}^r \ , \tag{3.7}$$

where $\Sigma = (\sigma_{jk})$. We introduce a notation $\Sigma^{(r)}$, which shows Hadamard product (elementwise product) (σ_{jk}^r).

Suppose that the items correspond to the variables of multivariate normal distribution, and category responses are discrete approximations of the values. Close consideration of MCA shows that the procedure is a process that selects the largest from the eigenvalues of $\Sigma^{(r)}$, $(r = 1, 2, \ldots)$. Suppose the eigenvalues of $\Sigma^{(r)}$ be $\lambda_1^{(r)} \geq \lambda_2^{(r)} \geq \cdots \geq \lambda_p^{(r)}$. Properties of Hadamard product of semipositive definite matrices lead to $\lambda_1^{(1)} \geq \lambda_1^{(2)} \geq \cdots$. These inequalities show that the first solution of MCA corresponds to the largest eigenvalue of $\Sigma^{(1)} = \Sigma$ except some degenerate cases. This MCA solution represents the first PCA components of the latent covariance structure. But the second solution may not correspond to the second PCA component of Σ. There are two possible cases. In one case, the solution corresponds to $\lambda_2^{(1)}$. And in another case, it corresponds to $\lambda_1^{(2)}$. In the former case, the solution represents the second PCA component of the latent normal covariance structure. In the latter case, the solution is a linear combination of quadratic functions $G_2(X_1), \ldots, G_2(X_p)$.

4. Nonlinear PCA based on MTV and MGV

PRINQUAL is the most popular implementation of PCAOS. It is able to use spline transformations for continuous variable. OS-MOD, which uses quasi-Newton projection method for optimization, is another implementation of PCAOS. Although motivations behind PCAOS is similar to MCA, PCAOS use the item-category hierarchical structure. With this feature, PCAOS can avoid the problems.

Suppose that variables $X_1, X_2, \ldots, X_p$ are distributed according to $N(\mathbf{0}, \Sigma)$. Then the covariance matrix of $G_r(X_1), \ldots, G_r(X_p)$ is $\Sigma^{(r)}$. The ith scale of MCA is expressed as

$$f_i(X_1, \ldots, X_p) = a_{i1}g_{i1}(X_1) + \cdots + a_{ip}g_{ip}(X_p), \qquad (4.1)$$

and they are orthogonal to each other. One reason of the inadequate solutions is that the transformation functions $\{g_{ij}\}$ depend on i. Therefore if $\{g_{ij}\}$ does not depend on i, we can solve a part of the problem. In this case, only $\{a_{ij}\}$ depend on i. To decide g_{*j}, we must introduce an estimation criterion. MTV and MGV are criteria for this decision.

Suppose that $X_1, \ldots, X_{p_1}$ are discrete and $X_{p_1+1}, \ldots, X_p$ are continuous. Let p be $p_1 + p_2$. And let $\{C_{j,1}, \ldots, C_{j,K_j}\}$ be the categories of X_j for $j = 1, \ldots, p_1$. Assigning scale values $\{w_{j,1}, \ldots, w_{j,K_j}\}$ to these categories, we can obtain p_1 variables

$$Y_j(\boldsymbol{w}) = \sum_{k=1}^{K_j} w_{j,k}\delta(j,k), \ (j = 1, \ldots, p_1), \qquad (4.2)$$

where $\delta(j, k)$ shows a dummy variable that indicates the category of the jth variable. If jth variable has category $C_{j,k}$, $\delta(j, k)$ is one. Otherwise it is zero. The vector $\boldsymbol{w}$ shows $(w_{j,k})$, $(j = 1, \ldots, p_1; k = 1, \ldots, K_j)$. For continuous variables, let

$$Y_j = \alpha_j + \beta_j X_j, \ (j = p_1 + 1, \ldots, p), \qquad (4.3)$$

where α_j and β_j are constants for standardization. Let the covariance matrix of $Y_1, \ldots, Y_p$ be $\boldsymbol{\Sigma}_{YY}(\boldsymbol{w})$. If $\boldsymbol{w}$ has appropriate values, then we can expect that the PCA components of $\boldsymbol{\Sigma}_{YY}(\boldsymbol{w})$ represent the latent structure of $X_1, \ldots, X_p$ well.

One criterion of MTV is defined as follows. Suppose that the dimension of analysis q is defined a priori. MTV is defined as maximizing the following under the constant variance conditions.

$$\theta = \lambda_1 + \cdots + \lambda_q , \qquad (4.4)$$

where $\lambda_1, \ldots, \lambda_p$ are the eigenvalues of $\boldsymbol{\Sigma}_{YY}(\boldsymbol{w})$ in descending order.

Another criterion of MGV is defined as maximizing the following under constant variance conditions.

$$\eta = -\log|\boldsymbol{\Sigma}_{YY}(\boldsymbol{w})| = -\sum_{j=1}^{p} \log \lambda_j , \qquad (4.5)$$

where $|\boldsymbol{\Sigma}_{YY}(\boldsymbol{w})|$ shows the matrix determinant of $\boldsymbol{\Sigma}_{YY}(\boldsymbol{w})$.

Theorems on Hadamard products of semi-positive definite matrices (Styan, 1973) leads to the following propositions (Otsu, 1993).

Proposition 1 :
Suppose a p dimensional normal distribution $N(X_1, \ldots, X_p|\boldsymbol{0}, \boldsymbol{\Sigma})$. Let the transformations of the variables by polynomials $\{g_j\}$ be $\{Y_j = g_j(X_j)\}$. The variances of $\{Y_j\}$ are constrained to one. And let the covariance matrix of $Y_j, (j =$

$1, \ldots, p)$ be Γ. The following inequality holds for these $g_j, (j = 1, \ldots, p)$.

$$\det \Gamma \geq \det \Sigma. \qquad (4.6)$$

Proposition 2 :
Let the eigenvalues of Σ be $\lambda_1 \geq \cdots \geq \lambda_p$, and the eigenvalues of Γ be $\xi_1 \geq \cdots \geq \xi_p$. The following inequalities hold.

$$\lambda_1 \geq \xi_1 \text{ and } \lambda_p \leq \xi_p . \qquad (4.7)$$

Proposition 2 assures the recovery of the latent PCA structure by MGV criterion, although under some non-regular conditions it may not be achieved. For example,

1. if Σ is singular, MGV is not appropriate,

2. if a variable X_j has zero correlations for other variables, we can not determine g_j uniquely.

Proposition 1 assures the PCA structure recovery by MTV with $q = 1$. Although the inequality $\sum_{j=1}^{q} \lambda_j \geq \sum_{j=1}^{q} \xi_j$ does not hold generally in case of $q \geq 2$, MTV works well for large eigenvalues (Otsu, 1993).

Figure 3 shows the category scores obtained by OSMOD with MGV criterion. The PCA structure of the latent distribution can be seen in the figure.

MGV criterion is essentially a scale invariant and it does not need predefined analysis dimension q. And numerical simulations verified its validity for small p (Otsu, 1993). On the other hand, MTV criterion is scale dependent and it gives unstable scale estimations if q is set too large.

Although MGV has these good properties, it may be inappropriate in applications for large datasets. The following are examples.

1. If the number of variable p or total number of categories are large, and the sample size is not sufficient, the determinant of the covariance matrix Γ may become singular.

2. Even if the covariance is not singular, instability of g_j may occur. Suppose that we must estimate optimal g_1 for the fixed $Y_2, \ldots, Y_p$. Let the number of categories in the first variable

be K_1. In this case, MGV optimization is canonical correlation analysis between K_1 dummy variables and $p-1$ continuous variables under standardizing restrictions of Y_1. If the covariance matrix of $Y_2, \ldots, Y_p$ is near singular, multicollinearity problems may occur. MGV criterion maximizes inter-dependence of Y_j's, therefore $Y_2, \ldots, Y_p$ tend to be linearly dependent.

3. If Y_1 has large correlation to one of $Y_2, \ldots, Y_p$, the determinant becomes near zero. In this case, other correlations have little effect on the criterion.

In general, scale invariance of a statistical criterion seems to be a preferable property, but from a different point of view it abandons some information. In the above cases, this causes inappropriate solutions.

5. Analysis of NLSY79 Data by OSMOD

In examples of PRINQUAL manual (SAS, 1985), most explanations are on nonlinear monotone transformations. This is also true in other PCAOS literature (Heiser and Meulman, 1995). But the most interesting examples of PCAOS are obtained from nominal measurement data. Monotone transformations rarely cause large changes in PCA structure for practical interpretations. In this section, we show an example of OSMOD analysis on NLSY (National Longitudinal Survey in Youth) data.

5.1 The Background of the Data

NLS is a series of large scale panel surveys by BLS (Bureau of Labor Statistics) of the United States. It has several research cohorts. NLSY79 is the project started in 1979. Its respondents are 12,686 young men and women born between 1957 and 1964. The respondents are consisted of three sub-sample groups. One is cross sectional samples (6,111 respondents). They are sampled to represent total U.S. population. Other sub-samples include supplemental samples (5,295 respondents). These sub-samples were oversampled on ethnic minorities and economically poor whites. The third group are military sub-samples (1,280 respondents). The number of supplemental sub-samples and military sub-samples were largely reduced up to now because of budget restrictions.

NLSY79 has had a high retention rate on large number of respondents. Detailed information on various aspects has been collected. It contains AFQT (armed force qualification test) points, high school grades, family conditions, marriage, delinquency and crime histories. One reason for the attention to NLSY was *The Bell Curve* by R. J. Herrnstein and C. Murray (hereafter we abbreviate it as TBC), which was published in 1994. They insisted on the importance of IQ for social achievements and strong influences of genetic properties on IQ. Based on these points, they argued that U.S. society is going to be stratified into social classes that represent their cognitive abilities i.e. IQ. Their arguments assume one-dimensional structure of cognitive abilities, which is called *g-factor*. Several psychologists who founded g-factor theory made large contributions to the development of psychometrics. C. S. Spearman, C. Burt, H. J. Eyzenck and A. Jensen are the prominent founders of g-factor theory. The argument of TBC is based on this work.

So far hundreds of thousands of copies of TBC have been published. Since the argument in TBC contained socially sensitive problems, it became a controversial book. There have been many arguments on TBC in psychology, sociology, statistics, and genetics. Jacoby and Glauberman (1995) and Fraster (1995) contain arguments against TBC. Fischer et al. (1996) showed critical reanalysis of NLSY data. Devlin et al. (1997) contains review articles in various areas. It also contains an extensive bibliography up to 1997. A summary of this book can be seen in Imrey (1999). The American Psychological Association started a task force, which is chaired by cognitive psychologist Ulric Neisser, for this problem. In Neisser et al. (1996), they reported on the current state of intelligence research. Beyond the problem of IQ and social achievement, the arguments in TBC and its criticisms gave important suggestions on the use of large social surveys.

5.2 Delinquency in NLSY79

The main data of NLSY79 were collected by interviews. Other data were collected by other methods. Achievement tests were performed by assembling respondents in local test places. The respondents answered self-report questionnaires on privacy sensitive items. Interviewers recorded the places where they met the respondents. For example, we can know whether the respondent was in a correctional institution or not.

Although the authors of TBC argued various problems, here we consider on delinquency and the correctional institution experience. Hereafter we call the latter *crime*. In TBC, data analysis based on NLSY79 were shown in the part 2. They analyzed only white male respondents for this issue to avoid arguments caused by ethnic differences. The analyses in TBC were performed as follows.

1. They composed a socioeconomic status index (SES) from family income, final grade of parents education, and parents occupational prestige index. SES was made by averaging the standardized values of these indices. If there were missing values, averages of valid data were used.

2. The authors used AFQT normal quantiles, SES, and Age as explanatory variables for various problems. Many analyses were performed for each sex.

3. Set a target variable for the analysis. Divorce rate, proportion of illegitimacy child, rate of professional workers, reception of official assistance, delinquency frequency, and weight of new born baby are the examples. The authors did logistic regression with the above explanatory variables.

4. Compare the standardized regression coefficient of SES and AFQT. In many cases, the coefficient of AFQT has larger values than SES. The authors claimed that these showed relative importance of native intelligence to growth environment.

The argument in the last half of TBC is rather speculative. Although part 2 is the most detailed empirical analysis in TBC, these analyses have the following problems.

1. Environmental factors are represented by an index SES. Its effect on social achievement should be more complicated.

2. The authors used AFQT as an index of innate cognitive ability. Citing famous twin studies (Bouchard et al., 1990), they claimed as follows.

   (a) IQ score (g-factor) is largely influenced by innate quality.

   (b) AFQT represents g-factor well.

But several longitudinal studies in a long period showed that IQ score had been considerably influenced by environmental factors (Flynn, 1987, 1999; Neisser ed., 1998). These findings suggest that the result "AFQT has stronger relations to the indices that represent social achievements than SES" does not directly lead to the claim "Innate quality is the most important for social achievement". Another problem is the quality differences between explanatory variables. AFQT measured by the respondent him/herself and had a rather reliable quality. But the variables used for composing SES were measurements on respondents' parents, and they had rather low qualities.

Fischer et al. (1996) and Manolakes (1997) did reanalyses of delinquency and crime data. In both articles, the authors claimed that the analyses in TBC underestimated environmental effects. These analyses showed that AFQT explanatory power was decreased by including more variables which represent environmental factors.

5.3 NLSY79 Analysis by OSMOD

In this equation we show an analysis of NLSY data using OSMOD. The analyses in both TBC and its criticisms use logistic regression. In large observational studies, selecting interactive variables for correct influence estimation is a difficult task. Here we try to extract the internal structure of the data to reconsider the regression results. We also investigate properties of MTV and MGV.

We used all men and women of three sub-samples for our analysis. Table 3 shows the variables for the analysis. Detailed results are shown in Otsu and Matsuo (to appear in Japanese). Some of these variables were used in TBC and its criticisms. Others were selected by us. We did not use parental occupation prestige, which was used for composing SES in TBC, because of rather large number of missing rates. We analyzed 7,025 samples that have valid responses to the variables. The data are composed of 4,434 cross sectional samples, 2,582 supplemental samples, and 9 military samples.

Table 4 shows the eigenvalues of the estimated correlation matrices. In MTV case, we set the analysis dimension q to be 3. There are slight differences in eigenvalues between two criteria. The configurations of the eigenvectors (factor loadings) estimated by MTV criterion are shown in Figure 4 and Figure 5. Interpretations of the eigenvectors for discrete variables depend on category scores. Table 5 shows the largest and smallest (negatively largest) categories for the discrete variables. The first PCA component represents family income and education. The SES in TBC corresponds to this component negatively. AFQT correlates strongly to the first component. These show that the analyses in both TBC and its criticisms tried to explain variables that have weak correlations to SES−AFQT component, and their explanatory variables are closely related to each other.

Now we examine the differences between MTV and MGV. Figure 6 shows the category scores of the variable WithMan that showed the largest scale differences between the two criteria. The vertical axis shows MGV, and the horizontal shows MTV. MGV scales tend to take more extreme values in this example. The estimated correlations are shown in Figure 7. The horizontal axis shows the mean of the two criteria. The vertical axis shows their differences MTV−MGV. In each case, 171 correlations are estimated. The correlations between continuous variables are the same. Therefore many points are on the horizontal axis. Many of other points are near the diagonal line. An isolated point in the right bottom region is the correlation between WithMan and WithWo. For this, MGV estimated a larger absolute correlation. These show that many larger absolute correlations were obtained by MTV. Many of the large differences were obtained by the correlations between WithMan and other discrete variables. MGV emphasized few correlations. These results show that MGV has problems for large dataset.

6. Conclusions

PCAOS is a good tool for exploring latent structure of large survey data. Though it is not based on sophisticated mathematical theory, its simplicity and moderated flexibility usually lead to good solutions. Theoretical considerations exhibited some good properties of MGV. Also numerical simulations with a small number of variables have shown its effectiveness. But in large survey data, It showed fragility. MTV with small dimensions is a good choice for exploring the structure of a large dataset with mixed measurement variables.

Fortran code of OSMOD and S-Plus/R interface function is available form the first author.

Table 3: NLSY Variables for the Analysis

No.	Symbol	Content	Measurement Level
1	Age	Age	Numerical
2	WithMan	With whom did R live at age 14 (adult man).	6 categories
3	WithWo	With whom did R live at age 14 (adult woman).	6 categories
4	MEdu	Highest grade completed by R's father in 1979	Numerical
5	FEdu	Highest grade completed by R's mother in 1979	Numerical
6	Sib	Number of siblings	Numerical
7	Reli	Present religious affiliation (1979)	10 categories
8	RelAtt	Frequency of R's religious attendance	Numerical
9	Race	R's Race	3 categories
10	Sex	R's Sex	2 categories
11	REdu	Highest grade completed as of May 1, 1992	Numerical
12	EmpStat	Employment status code	4 categories
13	LkJob	Global job satisfaction	5 categories
14	Delinq	Frequency of R's delinquency (constructed by us)	Numerical
15	Jail	Interviews in correctional institution	2 categories
16	Resi	Residence area (at age 14 and 1979)	4 categories
17	Pov	Under the poverty income criteria	3 categories
18	AFQT	AFQT89 score	Numerical
19	FmInc	Family income in 1979 and 1980	Numerical

(R means 'respondent'. Categories are treated as nominal levels.)

Table 4 : Estimated Eigenvalues by OSMOD

	1	2	3	4	5	6	7
Eigenvalues (MTV $q = 3$)	3.72	1.72	1.67	1.49	1.29	1.14	1.03
Minimum eigenvalue	0.27						
Accumulated ratios	0.20	0.29	0.37	0.45	0.52	0.58	0.63
Eigenvalues (MGV)	3.63	1.74	1.68	1.48	1.40	1.16	1.06
Minimum eigenvalue	0.27						
Accumulated ratios	0.19	0.28	0.37	0.45	0.52	0.58	0.64

Appendix

The computational details of OSMOD and some properties of PCA with nonlinear transformations are described in this appendix.

A.1 Numerical Procedures of OSMOD

OSMOD is composed of the following procedures.

1. Estimation for statistics. Means, frequencies, correlations, conditional means, and second order frequency tables (Burt table) are obtained.

2. Initial-value estimation for category weights.

3. Numerical optimization for the category weights. When we adopt

Table 5: Categories for Minimum and Maximum Scores

No.	Variable	Minimum	Maximum
2	WithMan	Relative	Father
3	WithWo	Relative	Stepmother
7	Reli	Baptist	Jewish
9	Race	White	Black
10	Sex	Male	Female
12	EmpStat	Employed	Out of labor force
13	LkJob	Missing	Like it very much
15	Jail	No	Yes
16	Resi	Rural-rural	Urban-urban
17	Pov	In poverty	Not in poverty

MTV criterion, this part composed of iterations of nonlinear optimization and eigenvalue decomposition. In nonlinear optimization, maximal weights are estimated for fixed eigenvector candidates. In MGV cases, eigenvalue decomposition is not used.

The first part is trivial. Other two parts are shown in the following.

A.1.1 Initialization of Category Weights : Initial-value estimation for ordered categories has little problem. Equivalent intervals,

$$w_{j,k+1} - w_{j,k} = w_{j,k+2} - w_{j,k+1}, \ (k = 1, \ldots, K_j - 2),$$

usually provide good estimators.

Nominal categories require more elaborated estimation. OSMOD uses a variant of canonical initialization that uses canonical correlation analysis (CCA) between the categories and other variables. This procedure was originally proposed by Tenenhause and Vachette (1977). The followings are the canonical initialization procedures for nominal scale categories.

1. Let T be the expanded set of the variables that is the union of dummy variables $\{C_{j,1}, \ldots, C_{j,K_j}\}$ where j for nominal category variables, $\{Y_j\}$ for ordered category variables with equivalent category intervals, and $\{Y_{p_1+1}, \ldots, Y_p\}$ for continuous variables.

2. Let $T(j)$ be $T - \{C_{j,1}, \ldots, C_{j,K_j}\}$ for j that is the variable with the nominal categories to be estimated.

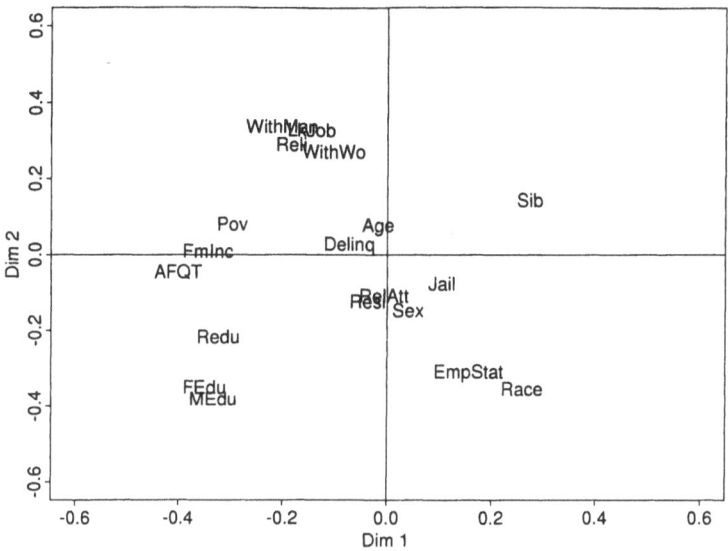

Figure 4: Eigenvectors by OSMOD, Dim.1 vs. Dim.2 (MTV, q=3)

3. Let $\{u_1, \ldots, u_m\}$ be the principal components of linear PCA
   on $T(j)$. CCA is performed between $\{C_{j,1}, \ldots, C_{j,K_j-1}\}$ and
   $\{u_1, \ldots, u_m\}$. Suppose the CCA correlation coefficients for the
   nominal categories be $\{a_1, \ldots, a_{K_j-1}\}$ and $\{a_{K_j} = 0\}$. According to their sorted order, set equivalent intervals for the categories. We can avoid extreme initial values by using equivalent
   intervals. This procedure depends on the expectation that the
   principal components of the expanded variables provide good
   approximations to the target principal components.

This procedure provides good initial values (Otsu and Saito, 1990).
The linear PCA is the critical part of this procedure. Without PCA
step, CCA coefficients do not provide good estimation.

A.1.2 Nonlinear Optimization : OSMOD uses iterative nonlinear optimization for estimating $(w_{j,k})$. Quasi-Newton projection method is
adopted for updating the estimates. In MTV cases, eigenvalue decomposition and nonlinear optimization are iteratively repeated. Quasi-
Newton method uses function values and its first order derivatives.
Hessians are estimated with iteratively computed derivatives. There
are some formulas for Hessian estimation. OSMOD adopted the most

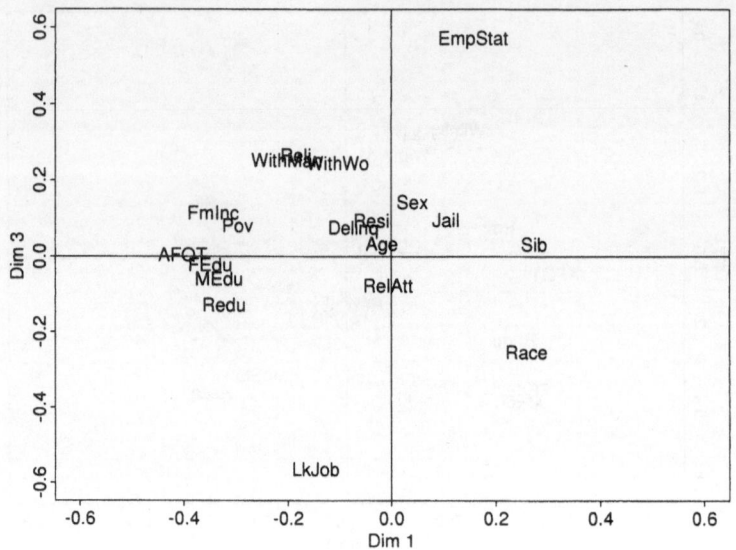

Figure 5: Eigenvectors by OSMOD, Dim.1 vs. Dim.3 (MTV, q=3)

popular formula that was proposed by Broyden, Fletcher, Goldfarb, and Shanno (Dennis and Moré, 1977; Konno and Yamashita, 1978). Let x_k be the estimated parameters at kth iteration, and g_k be the gradient vectors at this point. We write the difference of the two contiguous parameter vectors as $s_k = x_{k+1} - x_k$, and the difference of the two contiguous gradient vectors as $y_k = g_{k+1} - g_k$. The estimated Hessian matrix is described as B_k. The direction vector is defined as $d_k = -B_k^{-1} g_k$. BFGS formula for function *minimization* is defined as follows.

$$B_{k+1} = B_k + \frac{y_k y_k^T}{y_k^T s_k} - \frac{B_k d_k d_k^T B_k^T}{d_k^T B_k d_k}.$$

The equation $B_k d_k = -g_k$ leads to the following expression.

$$B_{k+1} = B_k + \frac{y_k y_k^T}{y_k^T s_k} + \frac{g_k g_k^T}{g_k^T d_k}.$$

Parameter updates are usually accompanied by rough linear search, therefore $s_k = \alpha_k d_k$ holds. For function *maximization*, the update formula is the same. But for these cases, B_k are negative definite. Let

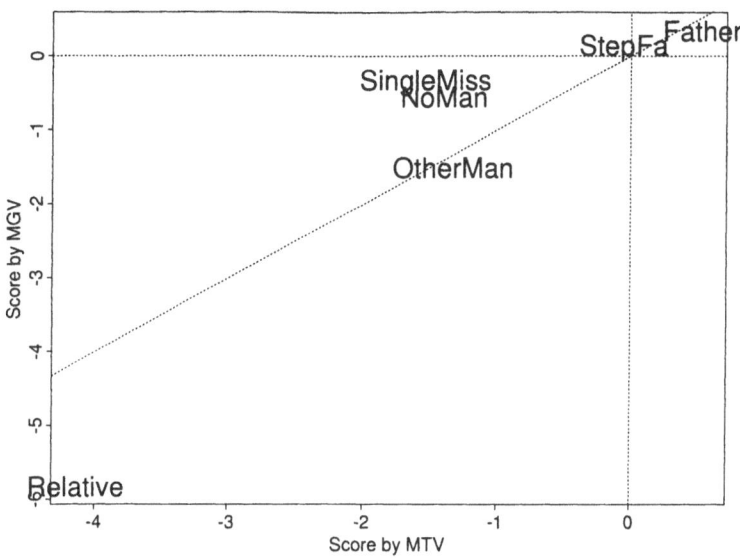

Figure 6: Category Scores for `WithMan` (Horizontal:MTV,
Vertical:MGV)

A_k be -1 times Hessian. Then

$$A_{k+1} = A_k - \frac{y_k y_k^T}{y_k^T s_k} - \frac{g_k g_k^T}{g_k^T d_k}, \tag{A.1}$$

where $A_k d_k = g_k$ holds.

In each update step, linear equation $A_k d_k = g_k$ must be solved. Although this requires decomposition of A_k, we can avoid decomposition by directly updating Cholesky factors. Suppose that $A_k = LDL^T$, where L is an under-triangular matrix with unit diagonals, and D is a diagonal matrix with positive elements. A BFGS update is composed of two one-rank updates, $LDL^T \pm uu^T$. A computational procedure that directly updates L and D is available (Gill, Murrayand Saunders, 1974, 1975; Tanabe, 1980). This methods ensure sign definiteness of A_k's. The new search direction d_k is obtained by solving $A_k d_k = g_k$. The solution vector d_k leads to the maximum point for quadratic function with exact A_k. Usually, rough linear search on this direction is necessary for reliable computation. OSMOD uses the following modi-

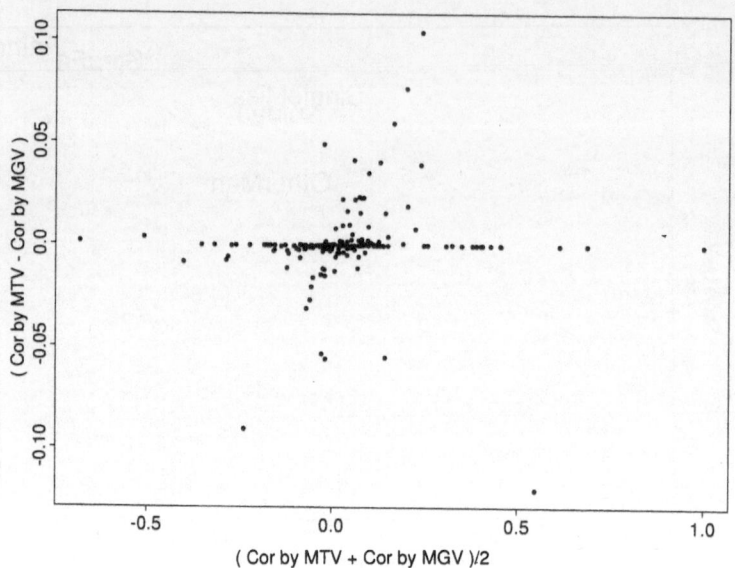

Figure 7: Estimated Correlations by Two Criteria Horizontal: Mean, Vertical: Difference

fied Wolfe's conditions for obtaining an admissible step size.

$$f(\boldsymbol{x}_k + \alpha \boldsymbol{d}_k) - f(\boldsymbol{x}) - \alpha \boldsymbol{d}_k^T \boldsymbol{g}_k \geq \varepsilon_1 \alpha \boldsymbol{d}_k^T \boldsymbol{g}_k \text{ and} \qquad (A.2)$$

$$f(\boldsymbol{x}_k + \alpha \boldsymbol{d}_k) - f(\boldsymbol{x}) \geq \varepsilon_2 \alpha \boldsymbol{d}_k^T \boldsymbol{g}_k, \qquad (A.3)$$

where we set $\varepsilon_1 = 10^{-4}$ and $\varepsilon_2 = 10^{-3}$. The former is to avoid too small steps, and the latter requires the sufficient gain per step size.

The category weights $(w_{j,k})$ are restricted to mean zero and variance one for each categorical variable. Therefore the number of free parameters for the jth variable is $K_j - 2$. OSMOD uses a simple reparametrization,

$$w_{j,k} = a_j + b_j t_{j,k}, \ t_{j,1} = 0, \text{ and } t_{j,K_j} = 1.$$

The two values a_j and b_j are defined as

$$a_j = \frac{-P_j}{\sqrt{NQ_j - P_j^2}} \text{ and } b_j = \frac{N}{\sqrt{NQ_j - P_j^2}},$$

where $P_j = \sum_{k=1}^{K_j} t_{j,k} n_{j,k}$, and $Q_j = \sum_{k=1}^{K_j} t_{j,k}^2 n_{j,k}$. The symbol $n_{j,k}$ shows the frequency of $C_{j,k}$.

For ordered categories, linear inequalities

$$t_{j,k} \leq t_{j,k+1}, \ (k = 1, \ldots, K_j - 1)$$

are imposed.

The correlation matrix $\boldsymbol{R}(w)$ of the variables (Y_j) is a function of t. The derivatives of the target functions are expressed as follows.

$$\frac{\partial \theta}{\partial t_{jl}} = \sum_{k=1}^{k_j} \frac{\partial \theta}{\partial w_{jk}} \frac{\partial w_{jk}}{\partial t_{jl}} \quad \text{and}$$

$$\frac{\partial \eta}{\partial t_{jl}} = \sum_{k=1}^{k_j} \frac{\partial \eta}{\partial w_{jk}} \frac{\partial w_{jk}}{\partial t_{jl}},$$

where $\theta = f_{MTV}$ and $\eta = f_{MGV}$. The partial derivatives of MTV and MGV with respect to (w_{jk}) are as follows.

$$\frac{\partial \theta}{\partial w_{jk}} = 2 \sum_{l=1, l \neq j}^{p} \sum_{a=1}^{m} \left(x_{ja} x_{la} \frac{\partial r_{jl}}{\partial w_{jk}} \right) + \sum_{a=1}^{m} x_{ja}^2 \frac{\partial r_{jj}}{\partial w_{jk}}$$

and

$$\frac{\partial \eta}{\partial w_{jk}} = 2 \sum_{l=1, l \neq j}^{p} \left(\frac{\partial \eta}{\partial r_{jl}} \frac{\partial r_{jl}}{\partial w_{jk}} \right) + \frac{\partial \eta}{\partial r_{jj}} \frac{\partial r_{jj}}{\partial w_{jk}},$$

where $\boldsymbol{R} = (r_{jl})$ is the estimated correlation matrix of the variables, and $\boldsymbol{X} = (x_{ja})$ is the estimated eigenvector matrix for the largest m eigenvalues. The derivative of η with respect to r_{jl} has a simple expression

$$\left(\frac{\partial \eta}{\partial r_{jl}} \right) = -(\boldsymbol{R}^{-1})^T = (-r^{lj}).$$

The derivatives $(\partial r_{jl}/\partial w_{jk})$ are expressed as follows. The symbol $\delta(j, k)$ shows the dummy variable that represents the kth category of the jth variable.

1. If lth variable has values of interval scale,

$$\left(\frac{\partial r_{jl}}{\partial w_{jk}} \right) = E_{sample}[\delta(j, k) Y_l], \ (l = p_1 + 1, \ldots, p).$$

2. If lth variable is categorical and l is not j,

$$\left(\frac{\partial r_{jl}}{\partial w_{jk}}\right) = \sum_{b=1}^{k_l} w_{lb} n_{jl,lb}/N.$$

3. For $l = j$,

$$\left(\frac{\partial r_{jj}}{\partial w_{jk}}\right) = 2w_{jk} n_{jk}/N,$$

where $n_{jl,lb}$ shows the response frequency to the cross category.

OSMOD adopts projection method for dealing with linear inequality constrains on ordered categories. Projection methods use reparametrization for new equality occurrences. If an estimated step size α lead to violation of inequalities, the step size is limited by the boundaries. The conditions that hold equality are called active. In our case, the procedure takes the following steps.

Suppose the following conditions are active on the jth variable.

$$0 = \quad t_{j,1} = \cdots = t_{j,k_1} \quad <$$
$$t_{j,k_1+1} = \cdots = t_{j,k_2} \quad <$$
$$\cdots$$
$$t_{j,k_{n_j-1}+1} = \cdots = t_{j,K_j} \quad = 1$$

The active conditions classify $\{t_{j,k}; k = 1, \ldots, K_j\}$ into equivalent n_j groups. Under the above equality constraints, the number of free parameters for the jth variable is $n_j - 2$. The new parameters are defined as follows.

$$\tilde{t}_{j,1} = \frac{1}{\sqrt{k_2 - k_1}} \sum_{k=k_1+1}^{k_2} t_{j,k},$$

$$\tilde{t}_{j,2} = \frac{1}{\sqrt{k_3 - k_2}} \sum_{k=k_2+1}^{k_3} t_{j,k},$$

$$\cdots$$

$$\tilde{t}_{j,n_j-2} = \frac{1}{\sqrt{k_{n_j-1} - k_{n_j-2}}} \sum_{k=k_{n_j-2}+1}^{k_{n_j-1}} t_{j,k}.$$

The inequalities on the new parameters are $0 \leq \tilde{t}_{j,1}/\sqrt{k_2 - k_1} \leq \tilde{t}_{j,2}/\sqrt{k_3 - k_2} \leq \cdots \leq \tilde{t}_{j,n_j-2}/\sqrt{k_{n_j-1} - k_{n_j-2}} \leq 1$. If new equations are activated, quasi-Newton optimization is performed on these new parameters.

If the process converges on the restricted subspace of $\tilde{t}$, Kuhn-Tucker conditions are examined. Suppose that t_0 is the convergence point on the restricted boundary. If the following inequalities hold, t_0 is a maximal point. For the first equivalent parameter block of the ordered item j,

$$-\sum_{k=h}^{k_1} \frac{\partial f}{\partial t_{j,k}} \geq 0, \text{ for } h = 2, \ldots, k_1.$$

For the intermediate lth parameter block $(1 < l < n_j - 2)$,

$$\sum_{k=k_{l-1}+1}^{h} \frac{\partial f}{\partial t_{j,k}} - \sum_{k=h+1}^{k_l} \frac{\partial f}{\partial t_{j,k}} \geq 0, \text{ for } h = k_{l-1} + 1, \ldots, k_l - 1.$$

And for the last parameter block,

$$\sum_{k=k_{n_j-1}+1}^{h} \frac{\partial f}{\partial t_{j,k}} \geq 0, \text{ for } h = k_{n_j-1} + 1, \ldots, k_{n_j} - 1,$$

where $k_{n_j} = K_j$. If an inequality does not hold, the corresponding active equality restriction is released. After reparametrization, quasi-Newton process is continued.

A.2 Proofs of the Propositions

We show the proof of two propositions and related properties of the nonlinear PCA model for normal distribution.

The following properties are known about Hadamard products of positive semidefinite matrices.

Theorem A1: (Schur, 1911; Styan, 1973, Theorem 3.4)

When A and B are $p \times p$ positive semidefinite matrices,

$$\text{ch}_p(A) \cdot b_{min} \leq \text{ch}_j(A * B) \leq \text{ch}_1(A) \cdot b_{max},$$

where $*$ denotes Hadamard product, and $\text{ch}_j(\cdot)$ denotes jth largest eigenvalue. Symbols b_{min} and b_{max} are the smallest and largest diagonal elements of B.

Proof of Theorem A1

Suppose that $B = TT^T$. Let t_j be the jth column of T, and x be a vertical vector of length p. The following formulas hold.

$$x^T(A * B)x = \sum_{j=1}^{p}(x * t_j)^T A(x * t_j) \leq \mathrm{ch}_1(A)\sum_{j=1}^{p}(x * t_j)^T(x * t_j)$$

$$= \mathrm{ch}_1(A)x^T(B * I_p)x \leq \mathrm{ch}_1(A)b_{max}x^T x. \quad \square$$

This proves the right inequality. The left inequality can be proved in the same way.

Lemma A1: (Mirsky, 1955; Styan, 1973, Lemma 3.2)

When A is positive semidefinite,

$$A_\alpha = A - \alpha e_1 e_1^T$$

is positive semidefinite. We write A_k as $(p - k) \times (p - k)$ submatrix of A without the first k rows and the first k columns. In the above formula, $\alpha = \det A/\det A_1$ when $\det A \neq 0$ and zero otherwise. We write e_1 as $(1, 0, \ldots, 0)^T$.

Proof of Lemma A1

If A is singular, A_α equals A. Otherwise, $\alpha = 1/e^T A^{-1}e$ holds. This leads to

$$A_\alpha A^{-1} A_\alpha^T = A - \alpha e_1 e_1^T = A_\alpha.$$

The left hand side shows the positive semidefiniteness of A_α. $\square$

Theorem A2 : (Oppenheim, 1930; Styan, 1973, Theorem 3.7)

When A and B are $p \times p$ positive semidefinite, the following holds.

$$\det(A * B) \geq \det A \times b_{11} \cdots b_{pp}.$$

Proof of Theorem A2

When A is singular or B has a zero diagonal element, the equality holds. Suppose A is nonsingular and B has no zero diagonal elements. Without losing generality, we can suppose that the diagonals of B are one. We write R as B that satisfies this condition. The inequality with this assumption is denoted as $\det(A * R) \geq \det A$. Theorem A1 and Lemma A1 lead to

$$\begin{aligned}0 &\leq \det(A_\alpha * R) = \det\{(A - e_1 e_1^T/a^{11}) * R\} \\ &= \det(A * R - e_1 e_1^T/a^{11}) = \det(A * R) - \det(A_1 * R_1)/a^{11}.\end{aligned}$$

Thus $\det(\boldsymbol{A} * \boldsymbol{R}) \geq \det(\boldsymbol{A}_1 * \boldsymbol{R}_1) \times \det\boldsymbol{A}/\det\boldsymbol{A}_1$ holds. Similar procedure shows $\det(\boldsymbol{A}_1 * \boldsymbol{R}_1) \geq \det(\boldsymbol{A}_2 * \boldsymbol{R}_2) \times \det\boldsymbol{A}_1/\det\boldsymbol{A}_2$. Finally we obtain the theorem because $\det(\boldsymbol{A}_{p-1} * \boldsymbol{R}_{p-1})/\det\boldsymbol{A}_{p-1} = 1$. $\square$

Suppose the following equations hold.

$$Y_j = \sum_r^q a_{jr} G_r(X_j), \ (j = 1, \ldots, p).$$

Let $\boldsymbol{\Sigma}$ be the covariance matrix of (X_j). We suppose the variances of $\{X_j\}$ are standardized. We write $\boldsymbol{\Sigma}^{(r)}$ as the rth power Hadamard product of $\boldsymbol{\Sigma}$ i.e.(σ_{ij}^r). Let the $p \times q$ dimensional coefficient matrix of Hermite polynomials be $\boldsymbol{A}$, and its rth column be $\boldsymbol{a}_r$. Then the covariance matrix of (Y_j), $\boldsymbol{\Gamma}(\boldsymbol{A})$, is denoted as follows.

$$\boldsymbol{\Gamma}(\boldsymbol{A}) = \sum_{r=1}^q (\boldsymbol{a}_r \boldsymbol{a}_r^T) * \boldsymbol{\Sigma}^{(r)} = \boldsymbol{\Sigma} * \left(\boldsymbol{a}_1 \boldsymbol{a}_1^T + \sum_{r=2}^q (\boldsymbol{a}_r \boldsymbol{a}_r^T) * \boldsymbol{\Sigma}^{(r-1)} \right). \quad \text{(A.4)}$$

When $\boldsymbol{\Sigma}$ and $\boldsymbol{\Gamma}$ are correlation matrices, the following holds.

$$\text{diag} \left(\boldsymbol{a}_1 \boldsymbol{a}_1^T + \sum_{r=2}^q (\boldsymbol{a}_r \boldsymbol{a}_r^T) * \boldsymbol{\Sigma}^{(r-1)} \right) = \boldsymbol{I}_p.$$

These expressions and theorem A2 lead to proposition 1. Using theorem A1, we can prove proposition 2.

A.3 Excessive PCA Dimensions for MTV

When we use MTV criterion, we must define PCA dimension m for the analysis. There are cases where overestimated m leads to indefinite category weights.

Suppose $\boldsymbol{\Sigma}$ be a correlation matrix that has a same off-diagonal positive elements σ. The eigenvalues of $\boldsymbol{\Sigma}^{(r)}$ are

$$\lambda_1^{(r)} = 1 + (p - 1)\sigma^r \text{and}$$

$$\lambda_2^{(r)} = \cdots = \lambda_p^{(r)} = 1 - \sigma^r.$$

When m is set to one, MTV criterion leads to the first PCA components of $\boldsymbol{\Sigma}$. If we assumed m was two, a problematic situation would occur.

MTV criterion is written as a function of $\boldsymbol{A}$ and $\boldsymbol{X}$,

$$\theta(\boldsymbol{X}, \boldsymbol{A}) = \sum_{r=1}^m \boldsymbol{x}_r^T \boldsymbol{\Gamma}(\boldsymbol{A}) \boldsymbol{x}_r,$$

where A is the coefficients matrix for Hermite polynomials, and $X_{p \times m}$ is a matrix of orthonormal vectors. We write x_r as the rth column of X.

We write v as $1/\sqrt{p}$, which is the eigenvector for the largest eigenvalues of $\Sigma^{(r)}$, $(r = 1, 2, \ldots)$, and P as vv^T. Let $A_r, (r = 1, \ldots)$ be diagonal matrices whose elements are $(a_{1r}, \ldots, a_{pr})$.

We write α_{ir}, β_{ir}, $(i = 1, 2; r = 1, \ldots)$ as

$$\alpha_{1r} = x_1^T A_r P A_r x_1, \qquad \alpha_{2r} = x_1^T A_r (I_p - P) A_r x_1,$$
$$\beta_{1r} = x_2^T A_r P A_r x_2, \quad \text{and} \quad \beta_{2r} = x_2^T A_r (I_p - P) A_r x_2.$$

Let $t_{1r} = \alpha_{1r} + \beta_{1r}$, and $t_{2r} = \alpha_{2r} + \beta_{2r}$ for $r = 1, 2, \ldots$. These satisfy $t_{1r} + t_{2r} = x_1^T A_r^2 x_1 + x_2^T A_r^2 x_2$ for $r = 1, 2, \ldots$. Orthonormality of X leads to the following inequalities.

$$
\begin{aligned}
t_{1r} &= v^T A_r (x_1 x_1^T + x_2 x_2^T) A_r v \\
&\leq v^T A_r^2 v = \operatorname{tr} A_r^2 / p.
\end{aligned}
$$

The equality holds if only if $A_r v$ is a linear combinations of x_1 and x_2. We write this condition as

$$A_r v \in \operatorname{Im}(x_1, x_2). \tag{A.5}$$

Applying the above inequality, the following is obtained.

$$
\begin{aligned}
\theta(X, A) &= \sum_{r=1}^m \left\{ \lambda_1^{(r)} \alpha_{r1} + \lambda_2^{(r)} \alpha_{r2} + \lambda_1^{(r)} \beta_{r1} + \lambda_2^{(r)} \beta_{r2} \right\} \\
&= \sum_{r=1}^m \left\{ \lambda_1^{(r)} t_{1r} + \lambda_2^{(r)} (x_1^T A_r^2 x_1 + x_2^T A_r^2 x_2 - t_{1r}) \right\} \\
&= \sum_{r=1}^m \left\{ p\sigma^r t_{1r} + (1 - \sigma^r)(x_1^T A_r^2 x_1 + x_2^T A_r^2 x_2) \right\} \\
&\leq p\sigma + 2(1 - \sigma) - \sum_{r \geq 2} \sum_{j=1}^p a_{jr}^2 (\sigma - \sigma^r)(1 - x_{j1}^2 - x_{j2}^2).
\end{aligned}
$$

The equality holds, if only if (A.5) holds for all r.

If we set $A_1 = I_p$ and $A_r = 0$ for $r \geq 2$, the upper limit $\theta = p\sigma + 2(1 - \sigma)$ is obtained. But this value is achieved by other (A_r) when $x_{j1}^2 + x_{j2}^2 = 1$ holds. Suppose $x_1 = v$ and

$$x_2 = \left(\sqrt{\frac{p-1}{p}}, \frac{-1}{\sqrt{p(p-1)}}, \ldots, \frac{-1}{\sqrt{p(p-1)}} \right)^T.$$

If $a_{ir} = a_{jr}$ holds for $i, j > 2$ and for all r, (A.5) holds. For $j \geq 2$, let $a_{j1} = 1$ and $a_{jr} = 0, (r \geq 2)$. Any $\{a_{1r} | r = 1, 2, \ldots\}$ that satisfy these conditions achieve the maximum θ. This leads to indefinite scale transformation of the first variable.

References

Bekker, P. and de Leeuw, J. (1988). Relation between variants of non-linear principal component analysis. In van Rijckevorsel and de Leeuw (1988), 1–31.

Benzécri, J. P. (1992). *Correspondence Analysis Handbook*. New York: Marcel Dekker.

Bouchard, Jr., T. J., Lykken, D. T., McGue, M., Segal, N. L., and Tellegen, A. (1990). Sources of human psychological differences: The Minnesota study of twins reared apart. *Science*, **250**, 223–228.

Dennis, Jr., J. E. and Moré, J. J. (1977). Quasi-Newton methods, motivation and theory. *S.I.A.M. Review*, **19**, 46–89.

Devlin, B., Fienberg, S. E., and Resnick, D. P. (eds.) (1997). *Intelligence, Genes, and Success*. New York: Springer.

Fischer, C. S. A. et al. (1996). *Inequality by Design*. Princeton NJ: Princeton University Press.

Flynn, J. R. (1987). Massive IQ gains in 14 nations: What IQ test really measure. *Psychological Bulletin*, **101**, 171–191.

Flynn, J. R. (1999). Searching for justice: The discovery of IQ gains over time. *American Psychologist*, **54**, 5–20.

Fraster, S., ed. (1995). *The Bell Curve Wars*. New York: Basic Books.

Gifi, A. (1990). *Nonlinear Multivariate Analysis*. Chichester: Wiley.

Gill, P. E., Golub, G. H., Murray, W. and Saunders, M. A. (1974). Methods for modifying matrix factorizations. *Mathematics of Computation*, **28**, 505–535.

Gill, P. E., Murray, W., and Saunders, M. A. (1975). Methods for computing and modifying the LDV factors of a matrix. *Mathematics of Computation*, **29**, 1051–1077.

Greenacre, M. J. (1984). *Theory and Applications of Correspondence Analysis*. London: Academic Press.

Greenacre, M. J. and Blasius, J., eds. (1994). *Correspondence Analysis in the Social Sciences: Recent Developments and Applications*. London: Academic Press.

Heiser, W. J. and Meulman, J. J. (1995). Nonlinear methods for the analysis of homogeneity and heterogeneity. *Recent Advances in Descriptive Multivariate Analysis*, edited by Krzanowski, W. J., 51–89, New York: Clarendon Press.

Herrnstein, R. J. and Murray, C. (1994). *The Bell Curve: Intelligence and Class Structure in American Life*. New York: The Free Press.

Hushimi, K. and Akai, I. (1981). *Orthogonal Function Systems*. Tokyo : Kyoritsu-Syuppan (in Japanese).

Imrey, P. B. (1999). Book reviews on Devlin et al. (1997). *Chance*, **12**, 7–11.

Jacoby, R. and Glauberman, N. (1995). *The Bell Curve Debate: History, Documents, Opinions*. New York: Random House.

Konno, H. and Yamashita, H. (1978). *Hi-senkei Keikaku-ho (Nonlinear Programming)*. Tokyo: Nikka-Giren (in Japanese).

Kuhfeld, W. F., Sarle, W. S., and Young, F. W. (1985). Methods of generating model estimates in the PRINQUAL macro, *SAS Users Group International Conference Proceedings: SUGI*, 962-971.

Kuhfeld, W. F. and Young, F. W. (1989). PRINCIPALS versus OSMOD: A comment on Saito and Otsu. *Psychometrika*, **54**, 755–756.

Lancaster, H. O. (1958). The structure of bivariate distributions. *Annals of Mathematical Statistics*, **29**, 719–736.

Lebart, L., Morineau, A., and Warwick, K. M. (1984). *Multivariate Descriptive Statistical Analysis: Correspondence Analysis and Related Techniques for Large Matrices*. New York: Wiley.

Manolakes, L. A. (1997). Cognitive ability, environmental factors, and crime. *Predicting frequent criminal activity*, edited by Devlin et al., 235–255.

Mirshky, L. (1955). *An Introduction to Linear Algebra*. Oxford University Press.

Neisser, U., Boodoo, G., Bouchard, Jr., T. J., Boykin, A. W., Brody, N., Ceci, S. J., Halpern, D. F., Loehlin, J. C., Perloff, R., Sternberg, R. J., and Urbina, S. (1996). Intelligence: Knowns and unknowns. *American Psychol-*

ogist, **51**, 77-101.

Neisser, U., ed. (1998). *The Rising Curve: Long-term Gains in IQ and Related Measures*. American Psychological Association.

Nishisato, S. (1994). *Elements of Dual Scaling: An Introduction to Practical Data Analysis*. L. Erlbaum Associates, N.J.: Hillsdale.

Okamoto, M. (1992). Artificial Data of Hayashi's third method of quantification, *The Japanese Journal of Behaviormetrics* (in Japanese), **19**(1), 75–82.

Okamoto, M. (1993). The Guttman effect of a linear trait in Hayashi's third method of quantification, *Mathematica Japonica*, **39**, 523–535.

Oppenheim, A. (1930). Inequalities connected with definite Hermitian forms, *J. London Math. Soc.*, **5**, 114-119.

Otsu, T. and Saito. T. (1990). The problem of local optimality with OS-MOD, *Psychometrika*, **55**, 517–518.

Otsu, T. (1990). Solutions of correspondence analysis with artificial data of typical patterns, *Behaviormetrika*, No.28, 37–48.

Otsu, T. (1993). OSMOD and its extensions: Investigations with Artificial Data, *The Japanese Journal of Behaviormetrics*, , **20**, 9–23. (in Japanese)

Otsu, T. and Matsuo, H. (to appear). An Analysis of NLSY79 Data by OSMOD. *Multivariate Analysis Practice Handbook*, edited by Yanai, H., et al., Tokyo: Asakura syoten (in Japanese).

Van Rijckevorsel, J. L. A. and de Leeuw J., eds. (1988). *Component and Correspondence Analysis*. Wiley.

Saito, T. and Otsu, T. (1988). A method of optimal scaling for multivariate ordinal data and its extensions. *Psychometrika*, **53**, 5–25.

Stuart, A. and Ord, J. K. (1987). *Kendall's Advanced Theory of Statistics, 5th ed. Vol.1*. London: Charles Griffin.

Styan, G. P. H. (1973). Hadamard products and multivariate statistical analysis, *Linear Algebra and its Applications*, **6**, 217–240.

Schur, J. (1911). Bemerkungen zur Thorie der beschränkten Bilinearformen mit unendlich vielen Veränderlichen. *Journal fuer die Reine und Angewandte Mathematik*, **140**, 1–28.

Tanabe, K. (1980). Hisenkei saisyo-zizyo-ho no algorithm (Algorithms for nonlinear least square methods). *Japanese Journal of Applied Statistics*, **9**, 119–140 (in Japanese).

Tenenhaus, M. and Vachette, J. L. (1977). PRINQUAL: Un programme d'Analyse en composantes principales d'un ensemble de variables nominales ou numeriques. *Les Cahiers de Recherche*, **68**. CESA, Jouy-en-Josas, France.

Tsuchiya, T. (1995). A quantification method for classification of variables. *The Japanese Journal of Behaviormetrics*, **22**(2), 95–109 (in Japanese).

Zagorsky, J. E., ed. (1997). *NLSY79 User's Guide 1997*. Columbus, Ohaio: Center for Human Resource Research. The Ohio State University.

Setting the Number of Clusters
in K-Means Clustering

Myung-Hoe Huh

Dept. of Statistics, Korea University
Anam-Dong 5-1, Seoul 136-701, Korea
stat420@mail.korea.ac.kr

Summary: K-means clustering is an efficient non-hierarchical clustering method, which became widely used in data mining. In applying the method, however, one needs to specify k, the number of clusters, *a priori*. In this short paper, we propose an exploratory procedure for setting k using Euclidean and/or Mahalanobis inter-point distances.

1. Introduction

One of the most important elements affecting the final output in K-means clustering is the specification of k, the number of clusters. Several tools are available for reasonable choice of k (cf. Milligan and Cooper 1985). One method is using the dendrogram from a hierarchical clustering, which is computationally more expensive. Besides such an *ad hoc* procedure, SAS, a popular statistical package program, offers some clustering criterions such as CCC (Sarle 1983, SAS Institute 1990). Another noteworthy procedures were proposed by Wong (1982), Wong and Lane (1983), Peck, Fisher and van Ness (1989), Rost (1995), and Jin (1999) among others.

The aim of this short paper is to propose an easily interpretable exploratory procedure for setting k. To that purpose, in Section 2, we develop the "MaxMin" algorithm which could be also useful for initial seeding. Numerical demonstrations are given on Monte Carlo simulation datasets in Section 2 and on real datasets in Section 3. More elaboratory procedures are given in Section 4, and final remarks on data mining are given in Section 5.

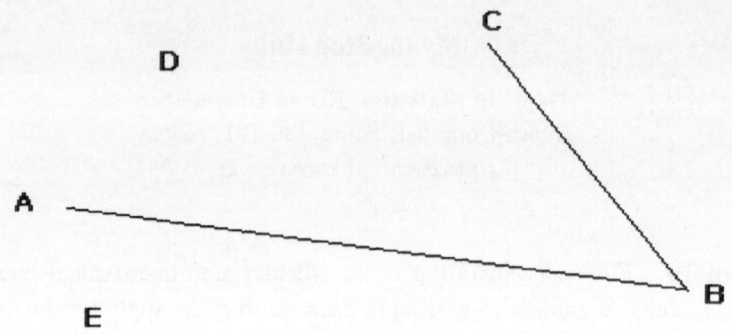

Figure 1: Illustration of MaxMin Algorithm with Five Observations

2. MaxMin Algorithm and Demonstration

The "MaxMin" algorithm we propose for exploratory choice of the number of clusters in K-means clustering is as follows:

Step 1: Compute the inter-point squared Euclidean distances for all pairs of observations (Hereafter "squared" will be omitted for brevity). Find the maximum among all inter-point distances. Denote corresponding observations as Seed 1 and Seed 2, and set $i = 2$.

Step 2: For each observation remained, compute the minimum distance to prechosen i Seeds. Find the maximum among all such minimum distances. Denote the corresponding new observation as Seed $i + 1$.

Step 3: Repeat Step 2 for $i = 3, 4, 5, \ldots$, until we have more than enough number of Seeds.

The rationale under MaxMin algorithm can be best illustrated with Figure 1, where there are five observations A, B, C, D and E. At Step 1, A and B are selected as Seeds 1 and 2, since they are furthest apart. At Step 2, there remain C, D and E. From C, B is the nearest seed with distance d(C,B). From D and E, A is the nearest with distances d(D,A) and d(E,A), respectively. Comparing these distances, we have d(C,B) $\geq$ d(D,A) and d(C,B) $\geq$ d(E,A). Hence C is selected as Seed 3. And so on.

As by-product, we believe that the MaxMin algorithm is useful for initial seeding procedure for K-means clustering even with the specified value of k, the number of clusters. So, the algorithm is used at the initial seeding stage in all K-means clustering of this study. But, we will not pursue the topic any more, since the initial seeding issue is somewhat off from the main scope of this research. See Sharma (1996, pp.202-207), SAS Institute (1990, p.825) and SPSS Inc. (1997, pp.448-449) for conventional initial seeding procedures in K-means clustering. Also, see Trejos, Murillo and Piza (1998) for applications of modern optimization techniques such as simulated annealing, genetic algorithm and tabu search to the clustering purpose in general.

For the demonstration of the MaxMin algorithm, we simulated Monte Carlo data as follows:

· Subsets of fifty observations are generated each from $N_7(\theta \mathbf{e}_j, I_7)$ for $j = 1, \ldots, 7$, the seven-dimensional normal distribution with mean vector $\theta \mathbf{e}_j$ and covariance matrix I_7, where

$$\mathbf{e}_1 = (1, 0, \ldots, 0)^t, \ldots, \mathbf{e}_7 = (0, 0, \ldots, 1)^t.$$

Thus, the centroids are at vertices of the simplex embedded in seven-dimensional Euclidean space.

· The artificial dataset of 350 observations is obtained by stacking seven subsets of 50 observations.

· Four sets are generated by varying the values of $\theta = 10, 5, 2.5, 0$. We call these datasets by Simplex 10, Simplex 5, Simplex 2.5, Simplex 0.

The results are shown in Figure 2. For the dataset Simplex 10 (Figure 2a), the maximum Euclidean distance drops abruptly after $k = 7$, indicating that the number of clusters is equal to seven. We may see a similar result for the dataset Simplex 5 (Figure 2b). But, for the dataset Simplex 2.5 (Figure 2c), we can hardly recognize any meaningful pattern. Dataset Simplex 0 (Figure 2d) provides the baseline reference for absolutely meaningless pattern.

3. Two Real Examples: Iris and Crab Data

First, we apply the MaxMin algorithm to well-known Fisher's iris data, which consists of 150 observations with four variables (x1: sepal length, x2: sepal width, x3: petal length, x4: petal width). In fact, each observation belongs to one of three species of iris (1: setosa, 2:

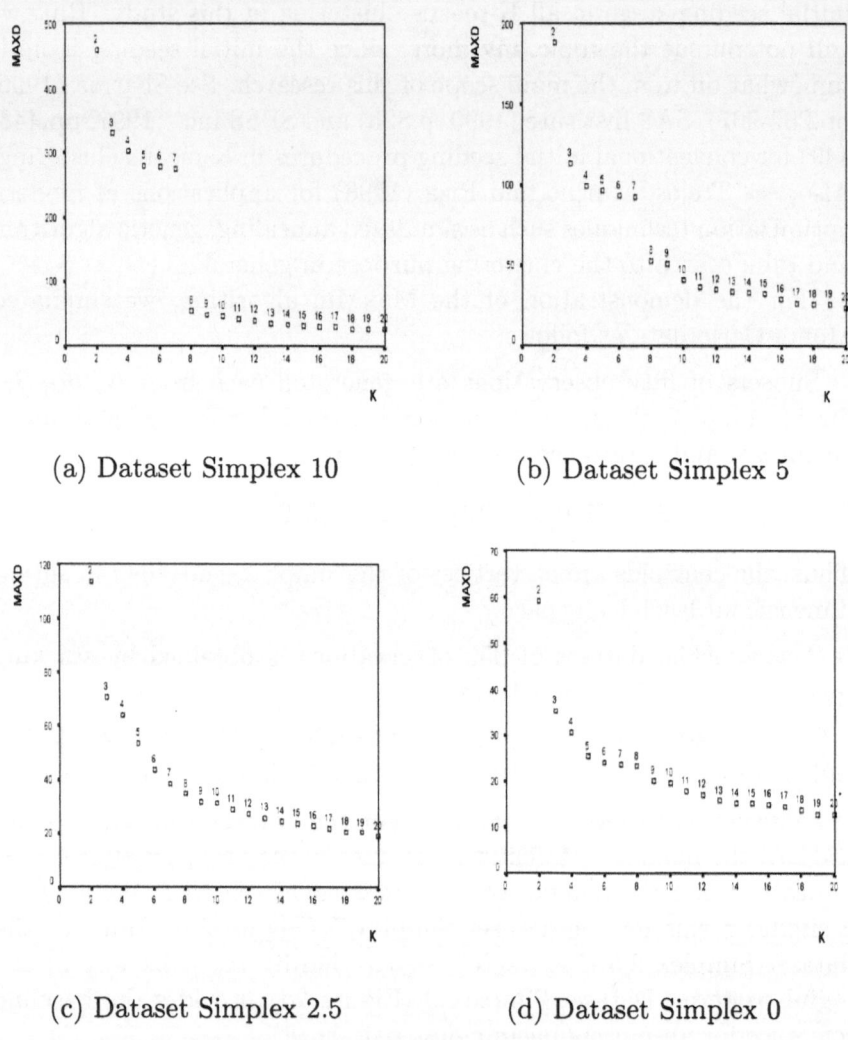

(a) Dataset Simplex 10   (b) Dataset Simplex 5

(c) Dataset Simplex 2.5   (d) Dataset Simplex 0

Figure 2: Sequence Plot of Maximum Euclidean Distances
for Simulated Datasets

versicolor, 3: verginica). But, for a while, we ignore the species for clustering purpose.

Sequence plot of maximum distances for standardized iris data is shown in Figure 3a. The plot suggests the existence of two, three or four clusters in the dataset.

Second, we apply our algorithm to Australian rock crab data (Campbell and Mahon 1974, Ripley 1996, p.13). Crab data consists of 200 observations with five variables (x1: frontal lip, x2: rear width, x3: midline length, x4: carapace width, x5: body depth). All variables are standardized before analyzed. Also, each observation is classified into one of four natural groups (1: blue male, 2: blue female, 3: orange male, 4: orange female), which will be ignored during clustering analysis.

One particular thing that can be observed in the dataset is high correlations between variables, ranging from 0.889 to 0.995. We may speculate that sampled rock crabs are varying with biological age spanning from early growing stage to well-grown stage. Obviously, the age is not a factor to be incorporated for identification of group membership. So, we performed a factor analysis to produce five factor scores, in which the first factor is considered as the "age". Hence, we use the second to the fifth factor scores as clustering variables.

Figure 3b is the sequence plot of maximum distances for crab data. The plot pattern indicates that there are two, four, or six clusters in the dataset.

Further elaborative method for the choice of k, the number of clusters, will be discussed in Section 4.

4. Further Elaboration

When the data can be considered as a mixture of k multivariate normal distributions with a common covariance matrix Σ, Euclidean distances may not be proper since the covariance structure is not incorporated. Hence, if we know Σ in advance, Mahalanobis distances could be adopted instead of Euclidean distances in the MaxMin Algorithm of Section 2.

Art, Gnanadesikan and Kettenring (1982) provides an estimate of Σ, without assuming the number of clusters k (SAS Institute 1990, PROC ACECLUS). It works very nicely for several cases, even though one needs to specify a critical constant, the proportion of observation

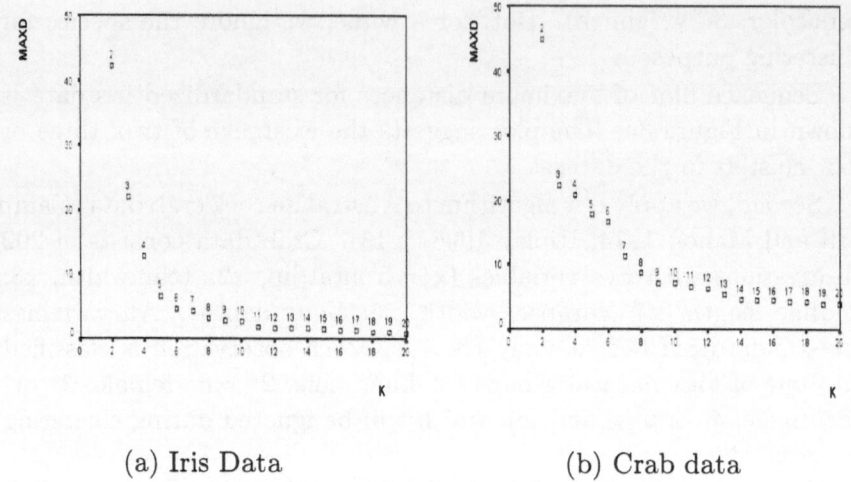

(a) Iris Data   (b) Crab data

Figure 3: Sequence Plot of Maximum Euclidean Distances
for Real Datasets

pairs to be included in the estimation of the within-cluster covariance matrix.

In this section, we propose another line of attack from a similar aspect as in Huh's (2000) "Double K-means Clustering":

0) Perform an ordinary K-means clustering with specified value k_0 for k, the number of clusters. The choice of k_0 could be guided by a sequence plot of maximum Euclidean distances of Sections 2 and 3.

1) Obtain an estimate S of Σ by pooling k_0 within-cluster covariance matrices.

2) Perform another K-means clustering with Mahalanobis distances. In other words, allocate observations to the nearest cluster centroid in the sense of Mahalanobis distance.

3) Repeat Steps 1 and 2 until the change in S, current estimate of Σ, is sufficiently small.

4) Make a sequence plot of maximum Mahalanobis distances with the final estimate S of common covariance matrix Σ.

If the maximum Mahalanobis distance drops off after some value k_1, then update the number of clusters from k_0 to k_1. When $k_0 = k_1$, we may have firm belief on the number of clusters.

As the first numerical example, we return to Iris data. In Section

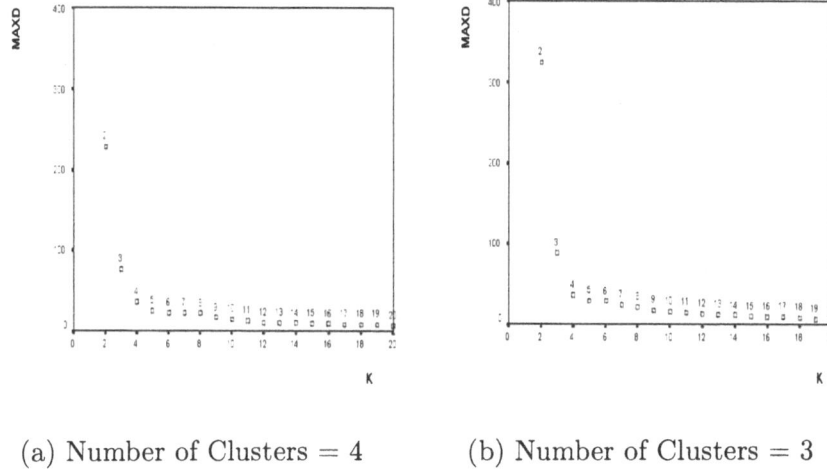

(a) Number of Clusters = 4   (b) Number of Clusters = 3

Figure 4: Sequence Plot of Maximum Mahalanobis Distances for Iris Data

3, we suspected that the number of clusters are equal to two, three or four. In cases that there are multiple choices, we may begin with the largest value for k_0. See Figure 4a for the sequence plot of maximum Mahalanobis distances drawn with $k_0 = 4$. The plot suggests $k_1 = 3$. Also, see Figure 4b drawn with $k_0 = 3$, which suggests $k_1 = 3$ too. Therefore, we may conclude that 'three' is the better choice than 'four' for the number of clusters k in Iris data.

As the second example, we continue to investigate the case of Crab data. In Section 3, we suspected that there are two, four or six clusters. Figure 5a is the sequence plot of maximum Mahalanobis distances drawn with $k_0 = 6$. It suggests $k_1 = 4$. Figure 5b is drawn with $k_0 = 4$, which suggests $k_1 = 4$ also. Hence, the conclusion is that 'four' is the better choice than 'six' for the number of clusters k in Crab data. Evidently, in both iris and crab data, two clusters are also a good choice.

Tables 1 and 2 are classification tables for clustering analysis of Iris and Crab data, respectively. In constructing tables based on Mahalanobis distances, K-means clusterings are repeatedly applied with updated estimates of covariance matrix. Compared with K-means clustering based on Euclidean distances, Mahalanobis version shows significantly improved classification accuracy in both datasets.

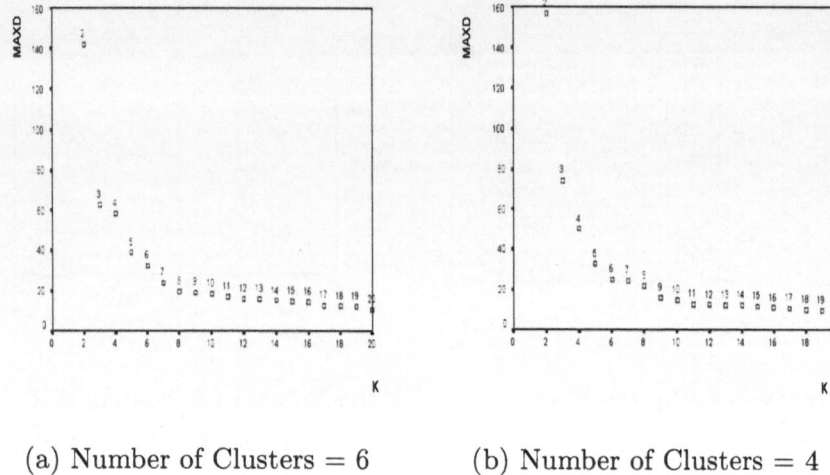

(a) Number of Clusters = 6   (b) Number of Clusters = 4

Figure 5: Sequence Plot of Maximum Mahalanobis Distances for
Crab Data

5. Concluding Remarks

K-means clustering is frequently adopted as a segmentation method
in data mining, where the data size is very large. For instance, we consider the case of which the number of observations n is 1,000,000 or
more. In such a case, the MaxMin algorithm could be inefficient, since
it needs the memory proportional to n^2. MaxMin algorithm is working
smoothly only with moderate value of n, say around 1,000. Hence, in
data mining context, MaxMin algorithm definitely needs modification.
We suggest the following.

First, select a random sample, so-called training sample, of moderate size from the full dataset. Second, apply MaxMin algorithm and
K-means clustering on the training sample to obtain various prelimi-
nary information such as the number of clusters k, the centroids and
the covariance structure. Third, allocate each observation in the full
dataset to the nearest centroid in Mahalanobis sense and update the
centroids immediately. At this stage, the allocation rule should be
flexible enough to allow several dissimilar observations or outliers to

Table 1: Classification Table for Clustering Analysis of Iris Data

(a) Based on Euclidean Distances

Variety	Cluster		
	1	2	3
1	0	1	49
2	13	37	0
3	42	8	0

(b) Based on Mahalanobis Distances

Variety	Cluster		
	1	2	3
1	0	0	50
2	0	48	0
3	46	4	0

Table 2: Classification Table for Clustering Analysis of Crab Data

(a) Based on Euclidean Distances

Group	Cluster			
	1	2	3	4
1	2	0	44	4
2	32	7	9	2
3	0	0	0	50
4	15	30	0	3

(b) Based on Mahalanobis Distances

Group	Cluster			
	1	2	3	4
1	8	0	37	5
2	49	0	0	1
3	0	0	0	50
4	1	46	0	3

form new clusters.

Cluster analysis can be formulated with mixture models, typically a mixture of multivariate normal distributions (Everitt and Dunn 1991, pp.113-120). At the expense of extensive computing, this approach may reveal more detailed clustering pattern in the dataset (McLachlan and Basford, 1988; Bensmail and Meulman, 1998).

Finally, we should mention that, to determine the number of clusters k, one needs to consider final use of the particular clustering or segmentation. Statistical choices of k including the one proposed in this study may give a suggestion only.

References

Art, D., Gnanadesikan, R., and Kettenring, J. R. (1982). Data-based metrics for cluster analysis, *Utilitas Mathematica*, **21A**, 75-99.

Bensmail, H. and Meulman, J. J. (1998). MCMC inference for model-based cluster analysis, *Advances in Data Science and Classification*, edited by Rizzi,A. and Vichi, M., Berlin: Springer.

Campbell, M. A. and Mahon, R. J. (1974). A multivariate study of variation in two species of rock crab of genus *Leptograpsus*, *Australian Journal of Zoology*, **22**, 417-425.

Everitt, B. S. and Dunn, G. (1991). *Applied Multivariate Data Analysis*. London: Edward Arnold.

Huh, Myung-Hoe (2000). Double K-means clustering, Unpublished manuscript (Submitted to *Korean Journal of Applied Statistics*, Written in Korean).

Jin, Seohoon (1999). A Study of the Partitioning Method for Cluster Analysis. Doctoral Thesis, Dept. of Statistics, Korea University. Seoul, Korea.

McLachlan, G. and Basford, K. (1988). *Mixture Models: Inference and Applications to Clustering*. New York: Macel Dekker.

Milligan, G. W. and Cooper, M. C. (1985). An examination of procedures for determining the number of clusters in a data set, *Psychometrika*, **50**, 159-179.

Peck, R., Fisher, L., and Van Ness, J. (1989). Approximate confidence intervals for the number of clusters, *Journal of the American Statistical Association*, **84**, 184-191.

Ripley, R. D. (1996). *Pattern Recognition and Neural Networks*. Cambridge: Cambridge University Press.

Rost, D. (1995). A simulation study of the weighted -means cluster procedure, *Journal of Statistical Computing and Simulation*, **53**, 51-63.

Sarle, W. S. (1983). Cubic Clustering Criterion, Technical Report A-108. SAS Institute, NC: Cary.

SAS Institute (1990). *SAS/STAT User's Guide* (Vol. 1), Version 6 Fourth Edition. SAS Institute, NC: Cary.

Sharma, S. (1996). *Applied Multivariate Techniques*. New York: Wiley.

SPSS Inc. (1997). SPSS 7.5 Statistical Algorithms. Chicago: SPSS Inc.

Trejos, J., Murillo, A., and Piza, E. (1998). Global stochastic optimization techniques applied to partitioning, *Advances in Data Science and Classification*, edited by Rizzi, A. and Vichi, M., Berlin: Springer.

Wong, M. A. (1982). A hybrid clustering method for identifying high-density clusters, *Journal of the American Statistical Association*, **77**, 841-847.

Wong, M. A., and Lane, T. (1983). A kth nearest neighbor clustering procedure, *Journal of the Royal Statistical Society* (Series B), **45**, 362-368.

Index